White Pine on the Saco River

An Oral History of River Driving in Southern Maine

Michael P. Chaney

NORTHEAST FOLKLORE

Volume XXIX   1990

Published annually by the Maine Folklife Center at the University of Maine, Orono, Maine.

Edward D. Ives, Editor

Pauleena M. MacDougall, Managing Editor

Maps and Diagrams by Steven Bicknell

Copyright © by The Maine Folklife Center, 1993

Printed by University of Maine Printing Services
Orono, Maine
1993

Northeast Folklore is the annual journal of the Maine Folklife Center, formerly published by the Northeast Folklore Society (1958–1991). Each year we hope to publish a single fresh collection of regional material or a comparative study, but we do not rule out the possibility of making a single volume out of several shorter collections or studies. Authors are invited to submit manuscripts for consideration to The Editor, *Northeast Folklore*, Maine Folklife Center, University of Maine, 5773 South Stevens Hall, Orono, Maine 04469-5773.

Front Cover Photograph: The cover photograph is identified in Frank C. Deering's own hand as "My first mill and crew—upper left: Charles Goodwin, John LeTarte, Fred Bunker and Albert Harvey. Upper right: Lewis Bean and Nathaniel Cousins. Lower: Eugene Williams and Albert Mack." Most probably this was the steam sawmill built 1893 on Springs Island, Biddeford, Maine and destroyed by fire February 6, 1913. A saw for trimming slabs and edgings appears in the right hand door. The Company was incorporated 1903 as J.G. Deering & Son—Frank as son, his father Joseph G. Deering having died in 1892. The photo was taken we presume between 1893 and 1902.

For my parents,
Margaret and Kenneth Chaney

Table of Contents

I. **GENERAL**

    1   **Introduction**

    8   **Maps**

    13   **Chapter One**   **An Overview of the Lumber Industry on the Saco River Valley**
            by Emerson W. Baker
            Dyer Library—York Institute Museum

II. **THE DEERING OPERATIONS REMEMBERED**

    19   **Chapter Two**   **Timber Cruising, Operating, and Scaling**

    32   **Chapter Three**   **The Saco River Drive**

    55   **Chapter Four**   **The Sawmill**

    73   **Conclusion**

    74   **Afterword**
            by Thomas M. Armstrong

    76   **Appendix One**
            Daily Journal of the Saco River Driving Company kept by Asa C. Cunningham on the
            1940 Log Drive

    80   **Appendix Two**
             Partial Listing of River Drivers, Truckers, Scalers, Teamsters, Portable Sawmill,
            and Woods Crews of J.G. Deering & Son, Saco River Driving Company, Diamond Match
            Company, and other companies on the Saco River and Tributaries, 1900 to 1960
            Compiled by Thomas M. Armstrong

    82   **Appendix Three**
            My Father was a River Driver
            by Clarence E. Brown

    83   **Appendix Four**
            A Short Story on Saco River White Pine
            by Frank C. Deering

    84   **Appendix Five**
            A Short History of J.G. Deering & Son
            by Joseph G. Deering

    85   **Notes**

    86   **Glossary**

    87   **List of People Interviewed**

    88   **Bibliography**

Special Acknowledgements

Publication of this book has been made possible by the generous support of the following businesses and institutions. The Maine Folklife Center staff and membership are extremely grateful to all of them:

Deering Lumber Inc. * Biddeford, Me.

Denison-Cannon Company * North Billerica, Ma.

Charles Diprizio & Sons, Inc. * Middleton, N.H.

Dyer Library-York Institute * Saco, Me.

Georgia Pacific Corporation * Portland, Me.

Gillies & Prittie Inc. * North Conway, N.H. and Scarborough, Me.

Thomas Hammond & Son * East Hiram, Me.

Hancock Lumber Company Inc. * Casco, Me.

H.G. Wood Industries Inc. * Bath, N.H.

Albert R. Lavalley Inc. * Sanford, Me.

Limington Lumber Company * East Baldwin, Me.

Lovell Lumber Co. Inc. * Lovell, Me.

George McQuesten Company Inc. * North Billerica, Ma.

Morton Lumber Inc. * Biddeford, Me.

Newman Lumber Co. Inc. * Woodsville, N.H.

New Hampshire Humanities Council * Concord, N.H.

Northeast Lumber Manufacturers Association Inc. * Cumberland, Me.

Pappys Inc. * Hiram, Me.

Robbins Lumber Inc. * Searsmont, Me.

Saunders Brothers * Westbrook, Me.

Timco Inc. * Center Barnstead N.H.

Winnipesaukee Lumber Company Inc. * Wolfeboro, N.H.

Introduction

This is a book about lumbering on the Saco River in New Hampshire and Maine from about 1920 to 1960. It tells the story of timber cruising, harvesting, river driving, and sawmill activity carried out by J.G. Deering & Son, now known as Deering Lumber Company. More than a dozen oral history interviews have been completed with people who worked for Deering or Diamond Match Company during this forty-year period. The sources, then, are first-person oral testimonies, and the book is very much in the style of previous *Northeast Folklore* volumes, especially *Argyle Boom* and *Suthin'*, and is intended to serve as a companion to them in documenting Maine's lumbering history.

Although the sources are primarily oral, in the course of examining business records we discovered the daily journal of the Saco River Driving Company kept by Asa C. Cunningham, Deering's woods and drive boss from 1926 to 1945. The journal documents daily events on the 1940 drive, from the first entry on April 8 to the last on August 13. We have used it frequently in the chapters describing the drive and have also printed it in its entirety in Appendix I. In addition, we have a rich photographic record of many aspects of the Saco Valley lumber operations.

In an earlier *Northeast Folklore* volume (XIX: 1978), Roger Mitchell told the story of his father's life as a woodsman. Some years later he described that book as follows: "From this personal saga emerges a body of motifs common to folklore in general. There are the outstanding bosses, the strong men, the pranksters, the cautionary tales of accidents that could have been avoided, the ability to withstand adversity. The unifying element is the working man's code: honest labor demands fair pay; work well done is time well spent; and honesty pays off in the long run."[1] The story of the Saco River contains all of these elements.

On the warm summer morning of August 14, 1980, my wife Laura and I went to the summer home of Joseph G. Deering in Biddeford Pool, Maine. An hour earlier, we had met Thomas Armstrong, then the owner of Deering Lumber Company. Tom had commissioned a series of oral history interviews which he hoped would preserve the history of the Saco River Drive and the lumbering enterprise which began as J.G. Deering & Son in 1866. With Tom along to make the introductions, we were met at the door and ushered immediately to Mr. Deering's study, a large room paneled in Saco River white pine—lovely wide boards running floor to ceiling—jointed and finished with precision and care. At the end of the room a huge bay window offered the best view of the calm waters of Biddeford Pool that one could hope for.

Mr. Deering, then eighty-six years old, got up from his desk in the center of the room and motioned for us to sit on the couch in front of the bay window. Walking across the room with a cane, he greeted the three of us and began talking to Tom Armstrong with enthusiasm about the prospect of Tom's son, C. D., taking a position with a lumber company. Mr. Deering spoke of how pleased he has always been about Tom buying the company in 1958. It was just wonderful news that another Armstrong generation would be in the lumber business. He immediately sat down in what he called his "chair that fits," with the telephone at one hand and stacks of *The Wall Street Journal* and *Barron's* at the other. On that day and at another session two weeks later, we tape-recorded Mr. Deering's memories of running J.G. Deering & Son and the Saco River Driving Corporation from his arrival in 1920 until he sold the retail lumber business to Tom Armstrong in 1958.

I took the photograph of Mr. Deering that appears in this book at the end of our second session. Unlike my experience with other casual portraits, this image is a particularly good likeness of the Joseph Deering that we interviewed. The interviews reveal a shrewd businessman who strongly believed in bartering—the cardinal rule in buying timber was never to take the other fellow's first offer. He put a high value on finding the right people to work for him, and keeping them. As he put it, "You've got to have people who know whether it's raining out or not." Joseph Deering also took great pride in running a business that had been in his family for generations.

The Deering family has been a part of the Saco-Biddeford business establishment since the mid-nineteenth century. Joseph G. Deering (1816–1892), considered a "prominent Pepperell Square grocer," built, with a partner, a sawmill on Springs Island in 1867. According to Roy Fairfield in his 1956 history of Saco—*Sand, Spindles and Steeples*—"at first, Deering supplied the capital and [*his partner*] Living Lane the managerial skill. But the former learned rapidly, and the partnership was dissolved in the early seventies."[2] The Deering family memory of how they got into the lumber business is more direct—it was a bad debt. Joseph Deering II put it this way: "Somebody persuaded my grandfather to loan him some money to build a sawmill beside a dam connecting Springs Island with Gooch's Island. As sometimes happens, the original amount loaned was not enough, and grandfather had to put in more and more until it finally proved that he had to take it over and start out in the lumber business... quite unintentionally."[3] By 1880, "the mill was shipping the greater portion of its three-quarter inch boards to Boston."[4] Moreover, the mill reportedly produced half as much lumber as the mills of the venerable Joseph Hobson, who dominated Saco River lumber production in the late nineteenth century.[5]

Joseph G. Deering, 1816-1892. Courtesy Dyer Library—York Institute, Saco, Maine.

In that same year, 1866, Joseph Deering's son, Frank Cutter Deering was born. The new lumbering enterprise was founded as J.G. Deering, and combined with income from his other businesses, Joseph Deering prospered. Four years later, he built the brick home at 371 Main Street, which is now the Dyer Library.[6]

Frank Deering left high school before graduating to work in the Deering businesses. Although his father intended that he manage a new grocery store in Biddeford, Frank was more interested in the lumber business. But in 1892, when Frank was twenty-six, his father died. Not only did he have the lumber business but also the grocery business, banking, real estate, and other investments to manage. He went on to become a director and eventually president of the York National Bank, a director of the Sweetser Children's Home and the Laurel Hill Cemetery. He lived at 371 Main Street until he died in 1939.

It is little wonder that Frank's son, Joseph Deering, born in 1894, developed such a reverence for family tradition. Joseph Deering was destined to carry on the family businesses in the twentieth century. He returned in 1920 from Yale and a stint in the Navy to help manage J.G. Deering & Son. As he relates in the annotated genealogy, *The Deering Family of Southern Maine*, after his father had gained controlling interest in the York National Bank in the early 1930's, "he ran the bank and gave up the lumber business, so he gave it to me. And he'd come over first thing in the morning and say, 'Don't do this, and don't do that,' and then he'd go back to the bank. It (the lumber business) was [*my father's*] until he died, no ifs, ands, [*or buts*] about it."[7]

But Frank Deering died, and Joseph assumed very much the same role as his father had played in the Saco-Biddeford community. As President of York National Bank, he would eventually see it merge with Canal National Bank; he would continue J.G. Deering & Son's lumber operations until 1958; and he would serve as President of Sweetser Children's Home, Laurel Hill Cemetery, and the Mutual Fire Insurance Company. After merging his remaining enterprises to form the Godfrey Company, Joseph Deering built and managed the Saco Valley Shopping Center, among many other business and charitable interests.

So it was three generations of the Deering family who were responsible for creating the lumbering enterprise which needed the Saco River to bring logs from the forests to the mill on Springs Island in Biddeford. The story of the Saco River drive is told only in part by Joseph Deering. There are the woods and drive bosses, the choppers, teamsters, river drivers, saw and planing mill workers, and office people. We not only need to know about the Deerings but also about the people who worked in the woods, on the river, and in the mill.

The series of oral history interviews which began with Joseph Deering in 1980 continued that year with twelve more sessions with nine men. Later, I interviewed three more people between 1987 and 1990. Because their quotes are scattered through the book, it is difficult to find just the right "spot" to introduce each person and provide some of the fieldwork context in which each person was interviewed. This introductory chapter, therefore, will acquaint the reader with the people telling the Saco River story.

My wife Laura and I saw Leo Bell of Fryeburg about a month after we interviewed Joseph Deering. Leo, at that time ninety-one years old, lived about three miles north of Fryeburg in a small white house located between the Odd Fellows Hall and the Fryeburg Town Hall on Route 5. The house sits in a field, but the woods are nearby. We talked with Leo at his kitchen table. He figured he began working on the upper part of the Saco River drive to Hiram as early as 1916 and continued until the last drive in 1943. First and foremost, Leo was a timber cruiser traveling the countryside buying woodlots, stumpage rights, or sawn logs. Like so many at this time in northern New England, he pieced together a living through farming, running the Fryeburg corn canning shop, buying timber, cutting timber, operating portable mills, and river driving. Leo Bell was one of Joseph Deering's primary woods foremen. He died in August 1986.

We saw Ken Blaney of Cornish about a week before we saw Leo Bell. Ken came to work for Joseph Deering in 1934 as a scaler—a person who determines how much

Frank Cutter Deering, 1866-1939. Courtesy Dyer Library—York Institute, Saco, Maine.

lumber is in a log. He was born near Woodstock, New Brunswick, in 1910, and five years later the family moved to Stillwater, Maine. He graduated from Old Town High School in 1929, worked for Jordan Lumber in Bangor, then made his way to the Saco River. Ken Blaney died in 1988.

Asa Cunningham, Ken Blaney's uncle, was Deering's woods and river drive boss from 1926 to 1945. Cunningham is something of a legend in this story—and he, like Leo Bell, worked for Deering over the long haul. Asa Cunningham's journal for the 1940 drive brings immediate detail to this book. Written on the day things happened, it provides striking perspectives to events the woodsmen and river drivers were remembering forty years later.

While Cunningham and Bell were the bosses, we found several men who shed light on what it was like to work in the woods or with a river crew. In his conclusion to *Joe Scott: The Woodsman-Songmaker*, Sandy Ives describes "moments of great integrity" that he has encountered, both in fieldwork and in his own life. The day I met Ed Burrill, a river driver from Cornish, is one I will always remember. He lives just south of Cornish proper, in a small cottage. When we showed up at their door, Ed and his wife had no idea who in the world we were. We were not able to reach them by phone because they didn't have one. The local postmaster had guided us to the Burrills' house. Ed is a stocky man with short white hair. His face, weathered by seventy-nine years in the outdoors, broke into a smile when I asked if we could talk with them on tape about the river drives. The house had a drainboard sink and a black kitchen woodstove used for heating both water and the four-room house. We sat for two hours in their kitchen/living room talking about the drives and looking at photographs. Ed first worked on the drive in 1929 and continued until the last drive in 1943. He described the geography of the river, the men he worked with, and told stories. My own "moment of great integrity", as we were leaving, was when Ed suddenly exclaimed, "I'd give *anything* to be right back in the middle of it again, on those logs in the river". The emotion does not come through on this printed page the way it did as I stood next to a man talking about a life he lived forty years earlier.

We met Clarence Brown at his farm in South Standish. Next to the farmhouse, located in the middle of a field, was one of the biggest trees that I have ever seen in the east. We parked our car next to a lot of other cars in back of the house. We found Clarence working on a chainsaw engine just outside the shed door. He had quite an array of other old chainsaw engines on the workbench, and it

Joseph G. Deering, II, 1894-1987. Photograph by the author, 1980.

looked as though we would do the interview right there, standing up, while he worked, until his daughter appealed to him to come inside and sit while we talked. Clarence agreed, but brought a chainsaw part in with him. He is the son of Frank Brown, Deering's drive boss until he died in 1923. Born in 1904, Clarence worked the drive itself for three years in the 1920's, and spent much of his time tending sluiceways in West Buxton and Limington. Having lost his father when he was a teenager just starting on the river, Clarence put his mind to writing "My Father was a River Driver," reproduced in the appendix. Clarence set about learning as much as he could about the river, the men, and the way things worked.

Bob Littlefield of Lovell went on his first river drive in 1920 at seventeen years old. He drove for about five years, tending sluiceways here and there, but he went the length of the river and remembers clearly when the older men let him pick off jams. Roy Smith is also from Lovell, and he worked in the woods and on the drives beginning in the 1930's. Chet Leonard told us about his work as a sawyer in the Fryeburg area, and Ralston Bennett told us about trucking lumber from Fryeburg to the mill in Biddeford. Ralph Morin provided details about the saw and planing mills. Phyllis Deschambeault, Irene Maher, Arlene Chappell, all of the Saco-Biddeford area, recounted the business of lumber production, wholesale and retail sales for the Deering Company. Alan Coker discussed working for Diamond Match Company in Biddeford from 1934 to 1979.

The remaining central figure we interviewed for the Deering story is Charlie Foran. Born in 1893, a few hundred feet from the lumber company, he was the yard and mill foreman of the Deering lumber manufacturing operation from 1933 to 1948, when sawmill production ceased. Charlie went to high school with Joseph Deering and was his lifelong friend and business associate. While Charlie worked for Joseph Deering, he made sure things were done Charlie's way, even if it meant telling Deering to stay out of the mill. It is Charlie Foran, with the help of Ralph Morin, who tells us how the sawmill and the planing or finishing mill operated.

Since most of the documentation of this book comes from oral interviews, it is important that the reader understand the methods of transcription and citation. Quotations preceded by an asterisk are taken verbatim from tape recordings; false starts, "uh's", and the like are the only things that have been eliminated without my indicating them. Omissions have been indicated by deletion marks, three (...) if the omission is less than a sentence, four if it is more. With these verbatim passages, material in brackets indicate the speaker's words were not absolutely clear, but this is my best guess as to what he or she was saying. In some places I have added my own words to clarify a passage. All my additions are italicized and bracketed. Oral material and attributed quotes not marked with an asterisk come from my field notes or my own recollection for which I cannot guarantee word for word accuracy. All of the tapes and accompanying transcripts are on file at the Northeast Archives of Folklore and Oral History at the University of Maine, Orono, Maine. Accession numbers and page numbers of transcripts are given wherever possible.

I have provided what I hope is an informative context of this oral history study. Tom Armstrong has been a driving force behind this book. Twelve years ago, he fortunately became acquainted with Sandy Ives and the work of the Northeast Archives of Folklore and Oral History. At that same time I was working for Sandy doing fieldwork on several projects and happened to be able to take on the Deering project.

The subject was not entirely new to me. A year earlier I had completed a National Endowment for the Humanities funded project on a photographer from Lincoln County, Maine, by the name of E. Joseph Leighton. He lived in my hometown of Sheepscot during the first half of the twentieth century. The NEH grant had enabled me to interview a dozen people and to create an exhibit with quotes from the interviews and marvelous enlargements of a quality that come only from five by seven inch glass plate negatives. Among other things, Joseph Leighton took lots of pictures of men working in the woods. Because of this project, I had met Sandy Ives as an undergraduate at the University of Maine. When writing the grant proposal to

Leo Bell 1889-1986 of Fryeburg, Maine on his woodlot in 1981. Note the pine stump—probably a mite younger than Leo himself. Courtesy of Betty Bell Baker, Fryeburg, Maine.

NEH, he helped, a lot. I will never forget taking care of the last detail of the proposal—Sandy's letter of support. He was on Prince Edward Island that summer, and of course I had to speak with him over the phone. His number was Eldon 5, ring 2. As I think back over the last twelve years, I marvel at what I (we) choose to remember. We chuckle and joke about things we remember from our childhood. But our memories reflect that which is most precious to us. That I remember Eldon 5, ring 2 is because meeting Sandy Ives was a turning point in my life. That I remember the summer of 1980 clearly is perhaps because I married the love of my life scarcely three weeks before we had our first meeting with Tom Armstrong and Joseph Deering. Laura and I did most of the interviews together, and we thought about them, compared our reactions, and contemplated together how different the world was just fifty years ago.

The interview project turned into a book project about three years ago. Although the writing is my own, you can be sure that Laura is in the fabric of the book. She, along with our children—Daniel, Philip, Eben, and Jessica—endured much while this book was written. In addition, Thomas Armstrong contributed greatly to the writing—describing technical terms, verifying how things worked, and locating photographs.

It has been a pleasure to meet a new colleague through this book, and I must thank Emerson Baker for a chapter that I as an outsider to the Saco Valley would have found difficult to write. Emerson Baker is director of the Dyer Library and York Institute (the library being the former Deering family home). He has provided a view of the Saco from his perspective as head of one of the major historical institutions in the Saco Valley. Shirley Santomassimo of Concord, New Hampshire typed and re-typed drafts of this book with speed, accuracy, and good humor. Thank you to Dick Lunt, Victor Konrad, and Joan Brooks who were supportive colleagues at the University during the time fieldwork was being done. James Garvin of the N.H. Division of Historical Resources, William Taylor of Plymouth State College, and Kathy Smith of the New Hampshire Humanities Council read early drafts of this book and have improved it immensely. Special thanks are due Charles G. Bickford, Executive Director of the New Hampshire Humanities Council, who supported this work by providing sabbatical leave needed to complete the project. I appreciate how he has kept after me to finish "that book" over the last three years.

The professional leave makes the New Hampshire Humanities Council a substantial contributor to this book, along with the financial support from sponsors listed at the beginning of the book.

A final acknowledgement goes to Sandy Ives. The content, scope, and approach of this book makes clear my intellectual debt to him, although I developed an interest in local community history much earlier in life, even before high school. My father showed me old cellar holes on the family property and read through daily journals of ancestral sea captains from voyages a hundred years ago. While my parents instilled in me a curiosity of what it was like to live in the past, Sandy Ives gave me a way of looking at history that did not try to link artificially the very real concerns of daily local life to some "great" or "significant" event(s). In too few courses with Sandy, I discovered the study of folklore and folklife. At last, I found out how to study people and communities from a cultural perspective. I found I could learn about daily life and work in my home town of Alna, or anywhere, by listening to people's stories, by studying their houses, walking their farms, and unfolding their quilts. Through historical photographs

Ken Blaney 1910-1988 of Cornish, Maine at Leo Bell's house in 1941. Courtesy of Betty Bell Baker, Fryeburg, Maine.

and stories that people could tell, I could reconstruct a community's history from *their* perspective, not from the historian's perspective, which may be commonly phrased, "How do these local occurrences fit into the larger social, economic and political picture?" Although I have asked similar questions in this book, I have been freed by the Ives approach to the history of life in the woods. This is a book composed of stories—narratives about the way things worked, about people, about fun, and about tragedy. When you come right down to it, stories about our families, their work, our work, and our communities—all these narratives combine to give us our own perspective on life. Without intending to be trite, I think we ought to take more time to listen to the stories.

Asa C. Cunningham 1875-1968 of Stillwater and Saco, Maine (far right). Deering's woods/river drive boss, 1926-1945. Courtesy of Leroy Smith, Lovell, Maine.

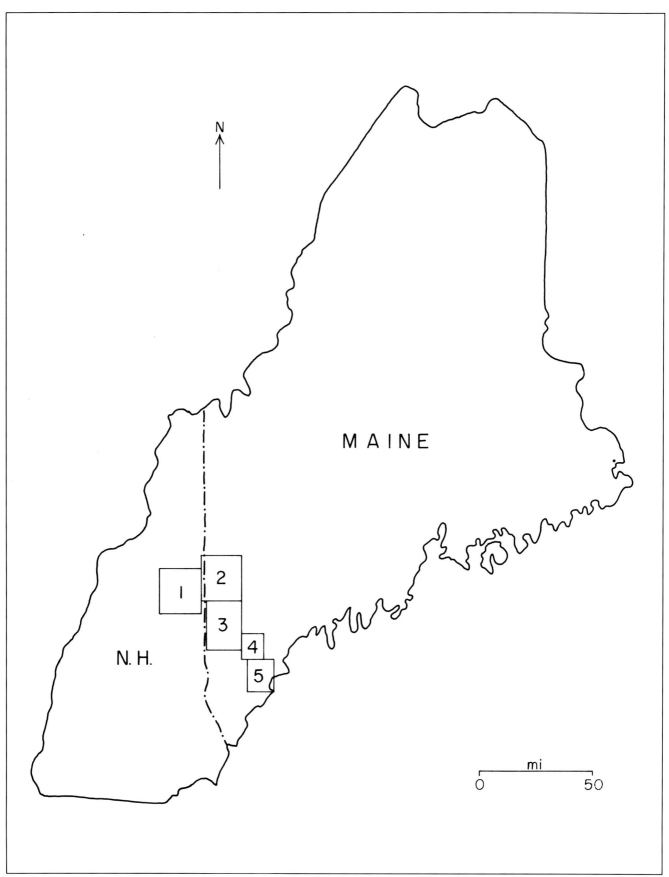

General map of the Saco River area. Numbered oblongs show areas covered by detail maps following.

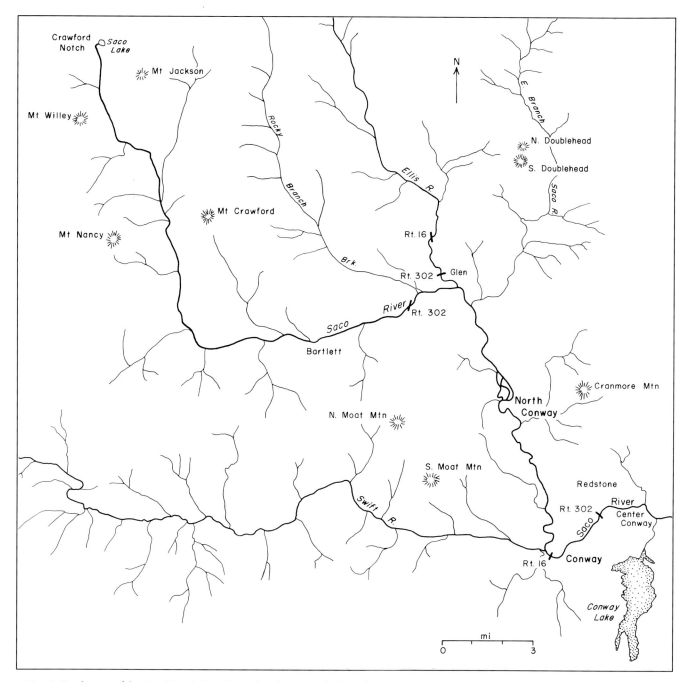

Map 1. *Headwaters of the Saco River in New Hampshire from Crawford Notch to Conway.*

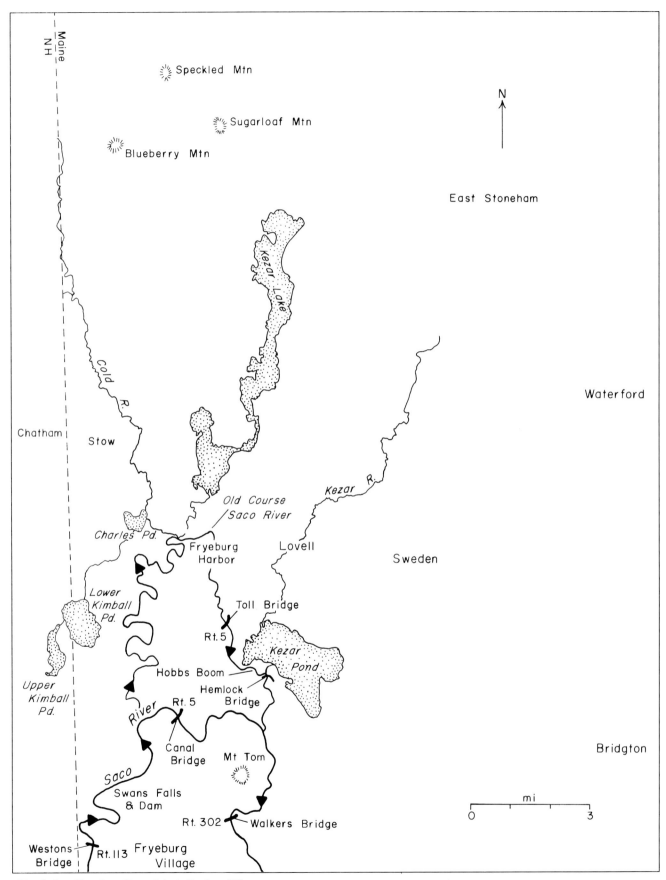

Map 2. Saco River from Kezar Lake to Fryeburg Village

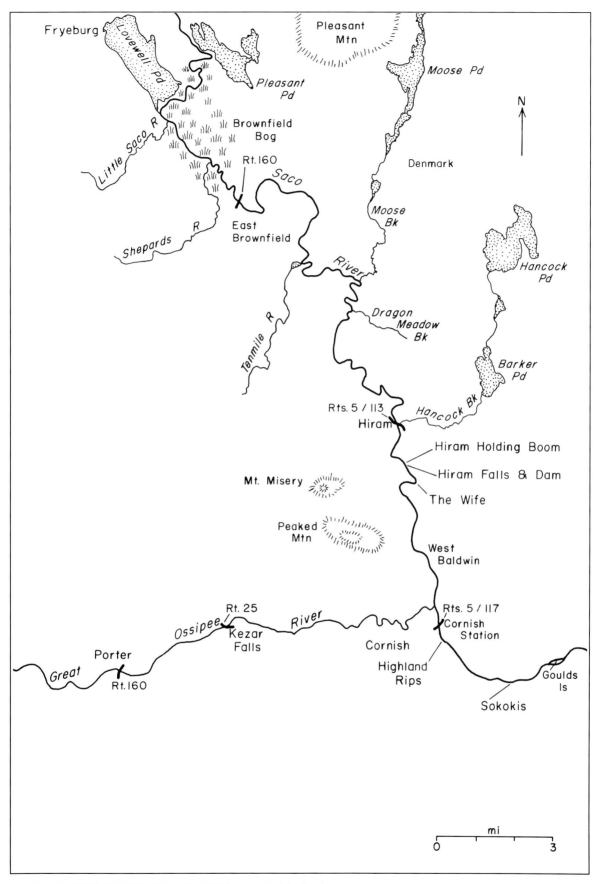

Map 3. Saco River from Lovewell Pond near Fryeburg to Goulds Island.

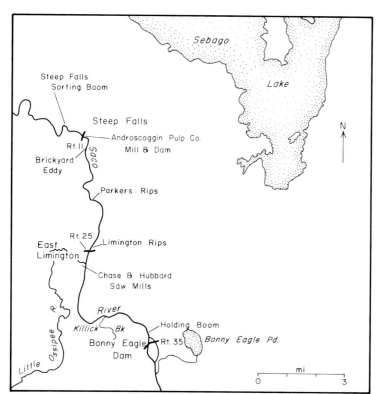

Map 4. Saco River from just below Goulds Island to Bonny Eagle Pond.

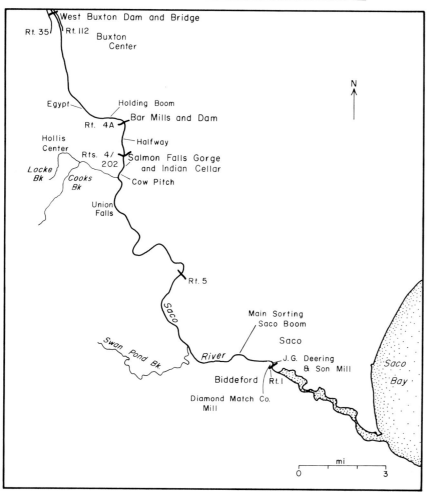

Map 5. Saco River from West Buxton Dam to Saco Bay. Note J.G. Deering & Son Mill and Diamond Match Co. Mill at Biddeford.

CHAPTER ONE

An Overview of the Lumber Industry of the Saco River

by Emerson W. Baker

The lumber industry has a long and rich heritage on the Saco River. While many of the first settlers were drawn to the mouth of the Saco for the rich fishing grounds, others eyed the prime stands of timberland. Settlers first moved to the lower reaches of the river in 1631, and by 1635 they were active in the lumber industry. For example, in that year the court of the Province of Maine administered the estate of Richard Williams, a "clapboard man," who possessed of 9,074 clapboards at the time of his death. Valued at the considerable sum of £ 164, the clapboards were apparently being shipped to Massachusetts Bay, used to build houses for the rapidly growing population.[1]

While clapboards could be cut by hand, before the logging industry could truly be successful, sawmills were necessary. The first sawmill privilege on the Saco was granted to Roger Spencer in a town meeting in 1653, but it is doubtful if he ever built the mill. He apparently only moved to Saco in 1658, and no surviving documents actually show him to be the owner of a mill on the river.

In January 1654 the town granted a sawmill privilege to John Davis. A blacksmith who had previously built sawmills in Portsmouth, Davis agreed to reduced prices of lumber for townsmen, and to set up a blacksmith shop in addition to the mill. The sawmill was built on the Biddeford side of the river, and according to one observer, it was "a little above the fall." By 1675 it was owned by Major William Phillips, a Charlestown merchant who lived on the west bank near the bottom of what is now Factory Island. At the time Phillip's residence was the furthest up the Saco, for settlement stopped at the falls. The sawmill and an adjacent grist mill were burned during an Indian raid on Major Phillip's house at the outbreak of King Philip's War in Maine in 1675.[2]

All settlers upriver from Biddeford Pool were forced to abandon their homes during the war, but peace came in 1678, and settlers quickly returned to their old properties. Two sawmills were a part of that rebuilding effort. In 1681 Benjamin Blackman, a Harvard-trained minister, purchased a sawmill built on the east side of the falls (in present-day Saco). Five years later, John Hill and his kinsman Francis Backus formed a partnership and constructed a dam and sawmill across Meetinghouse Creek (present-day Moors Brook in Biddeford), several hundred yards upstream from where it enters the Saco. Unfortunately

(Left to Right) Edward E. Burrill of Cornish (1901), Fred Hubert of Biddeford (1895-1967), Merle Cunningham, Asa's son, (1910-1973) and Stanley "Shorty" Harnett of Cornish (d. 1973). This photograph was taken on the drive in 1939 at the Limington side of the Steep Falls Bridge. Note calked boots and peaveys. Courtesy of Edward E. Burrill, Cornish, Maine.

these structures also were destroyed during a war with the Indians.

In 1688 hostilities resumed between the English settlers and the native peoples of Maine. Only several forts and a handful of settlers would dare remain before peace returned in 1714. At the time logging would have been a risky occupation but lumber men continued to work in the Saco Valley. For example about 1699 George Turfrey, a Boston merchant, built a mill just above the falls on the west bank. This mill would remain in operation until 1814, when it was swept away by a great spring freshet. While most boards cut from Turfrey's mill would have been exported, some were used locally in ship building which also continued during this time. Thus, not even the constant threat of Indian attack could completely halt the efforts to harvest the trees of the Saco.[3]

Soon after peace came in 1714, the logging industry of the Saco underwent tremendous growth. With the Indian threat gone it became safer and easier to cut logs from the forest, so lumber again became an important export of the region. Also as settlers returned to the region there was quite a strong local need for lumber. Mills rapidly developed on both sides of the falls to process the increasing amounts of locally cut timber. In 1719 Samuel Cole purchased twelve acres above the Turfrey Mill, where he built the Cole Mill. Other mills would also be built along the stretch of the riverbank in Biddeford which soon became known as the "mill brow." In 1741 Cole built a mill nearby on Gooch Island, in partnership with two Wells men, Thomas Wheelwright and Benjamin Gooch.[4]

While Samuel Cole was a leading force in the Biddeford milling industry, a Kittery merchant, William Pepperrell figured prominently in the rise of Saco's lumber industry. In two purchases in 1716 and 1717 Pepperrell purchased the land and mill privilege formerly held by Benjamin Blackman. The holdings included many acres of timberland directly upriver from the mill. Over the next several decades Pepperrell would add considerably to these holdings both near the fall as well as timberlands upriver. Sometimes he acted independently, but at other times he worked in joint ventures, providing the capital and letting others provide the milling expertise. For example, in 1716 Pepperrell purchased a large section of the original Saco patent. He then sold half of his holdings to Nathaniel Weare and Humphrey Scammon, Jr. These men then built a double sawmill and a dwelling house for the mill workers in partial payment for their share of the property.[5]

At the time of his death in 1759 Pepperrell's holdings on the east bank of the river contained about 5,500 acres, in addition to the sawmill, a gristmill, a wharf and a store. Pepperrell was an absentee landlord, only occasionally visiting Saco. He relied on his lands and his mills to provide export timber for the expanding trading business that he had inherited from his father. Pepperrell was one of

Frank E. Brown of South Standish (1874-1923). Deering's river drive boss from 1909-1923. Courtesy J.G. Deering & Son, Biddeford, Maine.

many New England merchants who participated in a world-wide trading pattern. From his wharf on the Saco River and his wharves and warehouses in Kittery he sent ships throughout the Atlantic Ocean. The West Indies, the Azores, Iberia, Newfoundland, Nova Scotia, and the southern English colonies were all destinations for the Pepperrell fleet, which numbered thirty-five vessels by the 1730s. The ships carried the raw materials of New England lumber, fish, and other foodstuffs, which were not as ready available in other areas. The trade with the Caribbean was particularly important for New England merchants in the seventeenth through mid-nineteenth centuries. In return for lumber and food, Caribbean merchants traded sugar and molasses to New Englanders, which could in turn be traded in England for export goods badly needed by New Englanders. Thus, in Pepperrell's time it was lumber and fish that were the basis of Maine's economy.[6]

William Pepperrell became the richest merchant of his day in northern New England. His wealth and position even led to his appointment as Commander-in-Chief of Massachusetts' 1745 expedition against the French fortress of Louisbourg. Pepperrell was knighted for the successful capture of this Cape Breton city. In 1762, three years after Sir William's death, the settlement on the east side of the Saco River separated from Biddeford. Because of Sir William's wealth, fame, and extensive holdings in the area, it is not surprising that the citizens named their new town Pepperrellborough.

The separation of Saco and Biddeford was just a sign of the many changes taking place in Maine in the 1760s. While Queen Anne's War had ended in 1713, a series of wars in the 1720s through 1750s had made it unsafe to venture too far into the Maine woods. As late as King George's War (1740-1748), isolated Indian attacks were made on the Saco settlements. It was not until 1759, when New France fell to British forces, that the frontier finally became truly safe for settlement. Settlers would pour into the Saco River Valley and the rest of Maine during the 1760s, and indeed most of the rest of the eighteenth-century. The boom rapidly pushed the line of settlement far inland and for a time made the Saco the center of the logging industry in Maine.[7]

The line of settlement, which had remained at Union Falls for decades, reached Fryeburg in 1763. Settlers were drawn to this distant spot by the fertile ground of the Fryeburg intervals; however within a few years the growing town would also be a center for logging activities in the upper Saco valley. In the latter 1760s and 1770s a series of settlements developed between Fryeburg and Union Falls. As farmers moved onto these forested lands, sawmills and the logging industry were not far behind. Indeed, they were an integral part of the settlement process. It was slow, difficult work clearing the densely wooded forests of interior Maine, but it was a necessary precursor to farming. The trees men felled were not simply an obstacle to farming, they were also a source of income. When a farmer cleared land, he could sell the lumber to a local sawmill or to the logging companies driving logs down river. During these times, logging would have made up a sizable percentage of income for many frontier farmers.

Local sawmills were also extremely important to a new settlement, for without them it was difficult for the community to grow. The very first settlers usually resided in crude log huts. Only when a local sawmill was set up could settlers begin constructing more permanent and more traditional forms of housing out of sawn boards and beams. Under such circumstances, it is understandable why towns would grant mill privileges to try to attract a sawmill. Since they were important to the growth of the community as well as the local economy, very often villages would develop around the sites of these early sawmills.

Settlement of any size only began in these townships in the 1760s. By 1790 over 2,100 people lived in Little Plantation (present-day Hollis and Dayton) and Buxton, and indeed Buxton's 1,508 inhabitants made it larger than either Saco or Biddeford. Many sawmills and much logging activity accompanied this growth. The first mill was built in Buxton in 1750, but by the end of the century many mills lined this stretch of the river.

The numerous falls of the Saco provided perfect spots to harness the power of the river for sawmills. Mills were built at Union Falls (Hopkins Mills), Salmon Falls, Bar Mills, Moderation (West Buxton), and Bonny Eagle Falls. In the nineteenth century many mills would be built in Limington, both on the Saco and the Little Ossipee, and the town would be recognized has having some of the best potential mill sites in the state. In addition to the Saco and the Ossipee, mills were built on some of their tributary streams and brooks. Indeed, the whole watershed from the ocean to its headwaters in New Hampshire bustled with logging and milling operations. In 1792 James Sullivan observed that the "great many mills and valuable works" on the Saco had "never destroyed or lessened the quantity of salmon in it," for "the mill dams do not extend across the river."[8]

While Sir William Pepperrell had dominated the logging industry in the early eighteenth century, Colonel Thomas Cutts symbolized the growth of the logging industry in the Saco River Valley in the late eighteenth and early nineteenth centuries. He was born in Kittery in 1736 to an old and well respected family. Thomas' father, Major Richard Cutts, represented Kittery in the Massachusetts legislature and also served as a county judge. Major Richard resided on Cutts Island on the family's comfortable estate.While young Thomas was born into a fairly wealthy family, as the ninth of ten children he did not inherit a large fortune. His family ties, however, helped place him at an early age in an apprenticeship with Sir William Pepperrell. The clerkship gave Cutts a chance to learn first hand both how to run a successful lumbering and mercantile business, and how to be a landed country gentleman. As his later life would prove, these were lessons Thomas learned well.

In 1758, when he was only twenty-two, Thomas left

Kittery and came to the Saco and set himself up in trading. He started with only a small sum of money lent him by his father; however, his keen business sense quickly allowed him to prosper. The next year he was able to buy a small plot of land on what was then known as Indian or Bonython Island (later known as Cutts Island and Factory Island) where he built a small house, with room for a store.

It was no coincidence that Cutts bought land on the island. He realized that the island held a great potential for commerce and industry. The island was well suited for wharfs and made an ideal spot for a ferry or bridge across the river. In addition, the falls on both sides of the island were good locations for sawmills and grist mills. Soon Cutts began to reap the benefits of his choice location. A bridge was built to Factory Island from Saco, and a new, shorter ferry began operating from Cutts's house to the Biddeford side. His mercantile business grew so that he soon had to make a larger store. Eventually Cutts would own all of Factory Island, and the mill privileges on the falls on both sides of the island. In addition to these holdings Cutts bought up extensive farms and tracts of land throughout the Saco River valley as well as 7,680 acres of land in Coos County, New Hampshire. At the time all of these timberlands were easy to acquire. The region was just opened for settlement, and large tracts could be purchased at a reasonable price. At the time of his death, Cutts owned a total of eighty-four pieces of real estate, which were valued in all at $96,000, a tremendous sum for this time. His extensive land holdings led some local residents to even claim that Cutts could travel to Canada by walking only on his property, and he could sleep in one of his own houses every night.[9]

Cutts purchased this land to continue to fuel his logging operations on the Saco. He built sawmills at the falls and then loaded his ships full of his lumber to participate in the West Indian trade. While Cutts was principally a merchant, his own involvement in the lumber industry meant that he could export his own lumber and effectively cut out the other loggers and the merchants who served them as middlemen. This practice clearly increased his profits.

The opening of the middle and upper Saco, and the large purchase of farms and timberlands by Cutts and other merchants brought a quick change in the nature of the logging industry of the Saco. In the seventeenth century when most of the river valley was forested, logs could easily be gathered from the areas directly adjacent to the sawmills of Biddeford and Saco. By the early eighteenth century lumberjacks had to go somewhat farther afield, but supplies could be obtained without going more than several miles up river. The lumberjacks preferred to stay close to the Saco and its tributaries, despite many acres of prime wood land in the interior. The reason was simple. The further from the river, or a major stream, the farther the log had to be hauled before it could be floated down river. The industry, which was already clearing land in Hollis and Little River in the 1740s and 1750s moved dramatically upriver in the later 1700s. Writing in 1830,

Irene and Ralph Morin, now of Sanbornville, New Hampshire, in June, 1991. Ralph worked in the Biddeford sawmill, planing mill, and retail department for forty-seven years, 1934-1981. Courtesy of Ralph Morin, Sanbornville, New Hampshire.

George Folsom observed that in the winter of 1772 "it is said that a few persons for the first time ascended the river as far as Fryeburg in quest of timber, and finding an abundance, turned the attention of millmen to that region for their future supply."[10] Thus, the first log drive down the Saco River from Fryeburg took place in the spring of 1772. In the 1780s, after the Revolutionary War ended, the practice resumed on a large scale and the Saco River was rapidly becoming the center of the logging industry in Maine.

While all the timber felling took place up river, a large percentage of milling occurred in Saco and Biddeford. Mills working up river produced boards largely for local consumption. Although some sawn boards were floated downstream for export, it was a difficult and time consuming task to transship boards around the many sizable falls of the river. Therefore, most of the boards shipped out of the region were cut right in Saco and Biddeford, the communities which not only enjoyed a tremendous set of falls for sawmills but also direct access to the sea and main shipping lanes. In the nineteenth century much of the Saco River's lumber went directly into the local ship building industry. In the 1820s and 1830s local newspapers often included advertisements from ship builders eager to purchase oak and pine. During the 1820s an average of 650 tons of ships were constructed in Saco and Biddeford each year.[11]

By the 1830s, however, the boom atmosphere was beginning to subside. Much attention and money turned instead to the logging operations of the Penobscot. The tremendous growth of Penobscot logging operations was making that river the leading lumber producer of the whole United States. The logging industry underwent a speculative boom in Maine in the early 1830s, with investors paying elevated prices for remote stands of timber. In 1835 the boom collapsed, leading to the ruin of many Mainers.

Even Ellis B. Usher, the king of Saco River logging operations, nearly went bankrupt during this collapse. A native of Medford, Massachusetts, at the age of twelve he

Bob Littlefield (1903) of Lovell and Clarence Brown (1904) of South Standish in Littlefield's kitchen in May, 1989. Courtesy of Bob Littlefield, Lovell, Maine.

and his brother moved to Hollis to seek their fortunes. By the time he was nineteen Usher had saved enough to purchase a farm and a share of ownership in a Buxton sawmill. In 1814 he was nearly ruined when a heavy flood wiped away his mills, dams, and $5,000 worth of logs, but he fought his way back to soon become the largest logging operator on the river. While the reverses of 1835 again threatened his business, he rebuilt it again, so that when he died in 1855 he was a wealthy man. The "rags to riches" story of Ellis Usher, complete with several booms and busts, seems in some ways typical of Saco's logging industry. Hard working men with good business sense could make a sizable profit in the business. Unfortunately a mill could be wiped out by a flood (or freshet), and frequent fluctuations in the price of lumber meant that financial reverses were a constant threat.[12]

By the 1830s the Saco logging industry not only suffered from a speculative collapse and competition from the Kennebec and Penobscot regions. It also vied for investment capital, and indeed for mill seats, with textile operations. In 1829 the first textile mill was completed on Factory Island, and while this wood framed structure soon burned to the ground, by 1850 a series of brick mills had developed along the falls on both sides of the river.

A gradual decline in prime lumber stands, combined with stiff competition from other Maine logging operations and the textile industry of the Saco meant that by 1850 the logging industry of the Saco was beginning to decline. By this time the most profitable prime stands which were close to rivers and streams were largely gone. A second growth of forest was growing and under harvest, but it was not as large or as profitable as the first. Although no longer dominant, for several decades the lumber industry of the Saco remained a driving force in the region. In 1850 there were about 240 sawmills, twenty clapboard machines, and 200 wood lathes on the upper reaches and tributaries of the Saco.[13]

In the second half of the nineteenth century, the logging industry was not only gradually declining in size and importance, it was also undergoing some significant structural changes. As railroads reached interior Maine a new transportation network allowed for easier shipment of boards and wood products. The development of effective steam engines also allowed for better sawmills, such as the steam powered mills built in Saco by Joseph Hobson. The markets for Maine lumber were changing as well. The urban markets of the East relied on other sources of timber, and the later nineteenth century also saw increasing urban use of brick and stone in construction. After mid century most of the once prosperous York County shipbuilders were also out of business. As a result of the loss of these markets for sawn boards, in the second half of the century more lumber was harvested and turned into finished wood products before it was shipped out of the Saco Valley. Wooden boxing became a principal product of the Saco. Boxes for the sugar industry were produced on the upper Saco and shipped out of Fryeburg by train. In Saco Joseph Hobson built a fortune largely upon wooden hogsheads for barrels used in the molasses and sugar industry. For example, in 1868 he produced 400,000 pairs of heads. Many were shipped directly from his Saco wharf to Cuba, while the rest went to Portland sugar refineries. At the time Hobson employed 150 men who manufactured $300,000 worth of products at his mills in the Somesville district of Saco and on Springs Island in Biddeford.[14]

Hobson was one of the leading lumberman of his time. The size of his operations and his kindly personality won him much respect, as well as his election in 1867 as Saco's first mayor. However, like others before him,

Charles and Anne Stebbins Foran of Saco, Maine. Charlie Foran (1893-1986) started with J.G. Deering & Son in 1920 and became yard and mill foreman in 1933. After the sawmill shut down, he left the company in 1950 to work for other lumber concerns, including Diamond Match, but eventually returned to work for Joseph Deering personally in matters associated with the Saco Valley Shopping Center. Courtesy of Patricia (Foran) Tibbetts.

Hobson was subject to the boom and bust environment of the lumber industry. Wiped out by the freshet of 1843, he rebounded to build the first gang mill on the Saco in 1851. During the Civil War the lumbering industry became depressed and Hobson went deeply into debt. Despite a post-war boom, these debts eventually forced him into bankruptcy. He rebounded in the 1870s, as his business soared to new heights. Unfortunately, he expanded too quickly, and in the mid 1880s a recession forced him into bankruptcy a final time. His estate was sold at public auction, and G.A. Crossman took over his business interests.[15]

In addition to Hobson's hogsheads, a variety of other wood products were turned out by Biddeford and Saco businesses during the Gilded Age. Samuel Shannon produced box shooks, as well as bats for baseball, the new sports craze that rose to prominence in post-civil war America. James Burbank's shop built soap boxes for factories in New Orleans. Charles Littlefield's mill on Lincoln Street in Saco made wheels and woodwork for local carriage factories. Smaller concerns such as the Eureka match company, and A.P. Moody's factory for doors, sashes and blinds, also waxed and waned during this era.[16]

The last big cut in the Saco Valley was in 1877 when 35 million feet of lumber were cut and driven into Biddeford and Saco. At this point the yearly cut was still being replaced by the yearly growth of trees. However, the industry rapidly declined after this. The cut in 1889 brought in 17.5 million feet, only half of the 1877 total.[17]

The lumber industry of the Saco had traditionally been dominated by one or two men and their companies, and the later nineteenth-century was no exception. The earlier generations had Thomas Cutts and Ellis Usher, and the mid to late nineteenth century had Joseph Hobson and Samuel Hamilton, a Waterboro-native who settled in Saco and bought a sawmill on Gooch Island after the Civil War. Hamilton became a somewhat legendary figure, for not only was he a successful merchant lumberman, but he also filled the traditional "rough and tumble" image of the lumberjack. A giant of a man, with unquestioned courage and a glib tongue for telling stories, Hamilton was an expert cant hook wielder, and the champion wrestler of the Saco Valley.[18]

Joseph Godfrey Deering was Hamilton's contemporary and fellow Waterboro native. Deering moved to Saco where he became a prosperous Pepperell Square grocer. He and his brother than branched out into the construction industry, building homes for the growing population of mill workers. From the construction industry Deering made a move into sawmilling, largely as a result of a bad debt. In 1867 he built a mill on Spring Island in partnership with Living Lane, an experienced lumberman. By the early 1870s Deering had learned enough from Lane to dissolve the partnership and go it alone. By the end of the 1870s this newcomer to the industry already produced half as many sawn boards as Joseph Hobson. By 1889 Deering's business had grown to an average of 6.5 million board feet a year, making him a dominant force in the valley's lumber industry. Yet it was an industry in decline, with the cut significantly reduced from the 1860s and 1870s. By the 1880s Joseph's son, Frank Cutter Deering had entered the firm, necessitating a change in the firm's name to J.G. Deering and Son. Eventually Frank's son, another Joseph G. Deering, would enter the firm and he would continue to run J.G. Deering during and after the last log drives of the Saco which are recounted in this volume.[19]

This introduction has focused on those merchants who led the industry, for it is those men who have left the most historical records behind from which to piece together an historical pattern. Fortunately, the oral testimony that follows provides the perspectives of all levels of employees in the lumbering business, including much that does not survive in other historical records. Yet, clearly the lumbering industry has affected the lives of most residents of Saco River Valley for the past 350 years, even if they were not employed in it. Many people have worked in related industries such as ship building and ship masting. Others have relied on Saco lumber to build the barrels, boxes, and crates used to export local products ranging from fish to textile machinery. Even farmers clearing fields and tending woodlots came to rely on lumbering as a part of their income.

The whole settlement pattern of the Saco River is much indebted to the lumber industry. Places such as Goodwin's Mills and Clark's Mills owe their existence (and indeed their names) to long forgotten saw mills. Other areas upriver saw less development as they continued to be held as timberlands. Saco and Biddeford grew into major commercial centers in the late eighteenth and early nineteenth centuries because they were the terminus for log drives in the Saco River watershed. The growth of the lumber industry would even affect town government. In 1762 at the Saco (then known as Pepperrellborough) town meeting, attendees appointed two surveyors of lumber, but in 1802 the town appointed twenty surveyors of lumber, five surveyors of logs, six measurers of wood and bark, and other officials to examine staves, shingles and clapboards.

By the mid-nineteenth century textile manufacturing had become the predominant force in Saco and Biddeford, but lumber would remain an important industry in the twin cities and throughout the valley. The Deerings ran the last drive down the river in 1943. Changing times forced changes in the industry, and brought different business opportunities. J.G. Deering and Son is still in existence, although it is now owned by the Armstrong family. Today the Deering family's principal commercial interest in the area is the Saco Valley Shopping Center, which occupies the site of the former Deering lumber yard in Saco. It is an interesting note to the story of the Saco River Valley that after the last log drive down the Saco the Deering's would become involved in general merchandizing, for most of the leading lumbermen of the Saco—from Sir William Pepperrell to the first Joseph G. Deering—had been merchants who had expanded their interests to include the logging industry. In this way, perhaps, the story has come full circle.

CHAPTER TWO
Timber Cruising, Operating Timber Lots, and Scaling

Timber Cruising

The identification and procurement of timber in the Saco River Valley was an ongoing and scattered affair. The success of a company's buying often hinged on the personal friendship and trust built over the years between buyer and seller. Although J.G. Deering & Son owned significant tracts of timberland in Lovell and Sweden, the company also bought stumpage (standing trees) from farmers and land owners all along the Saco River Valley and elsewhere. A third source of saw timber was logs cut and furnished by farmers, landowners, and independent loggers. In the Saco Valley the more desirable species for Deering and most lumbermen were white pine and spruce, but they also sought lower value Norway pine and hemlock logs. Oak ship and saw timber, oak and Norway pine piling, birch and maple logs and bolt wood were valuable but cut in lesser quantities.

The alert timber cruiser (buyer) was always on the lookout for these species. He needed also to deal with the smaller and coarser grades of pine that inevitably developed on any wood lot. Large quantities of poor quality were sawn "live" or "round edge" (bark left on both edges) for the box board and crating markets. Although that market is minimal today, the pulpwood market uses low grade softwoods in great volume. In addition, the timber buyer bought large quantities of birch, beech and maple hardwoods for charcoal and cord (fire) wood. The timber cruiser exercised his skill and knowledge best in the purchase of stumpage—walking a woodlot to determine its volume and quality. Leo Bell, on many lots, measured each tree with a tree caliper—calling sizes to his brother Merton Bell. The experienced cruiser could estimate volume very closely by counting trees in a small plot and estimating height (two, three or four sixteen foot logs to a tree). Merchantable pine stands might run ten to fifteen thousand board feet per acre, twenty thousand or more on choice lots. If the timber buyer and stumpage owner could then agree—the sale was made—often with a handshake. "Deering (or the Diamond Match Co.) was to have the lumber"....as the saying went. Purchase might be by the thousand board feet (MBF) or by lump sum—perhaps all the "black growth" (conifers) or "all standing wood over eight inch diameter"—or whatever terms were agreed on. Stumpage sales often allowed cutting rights of three to five years or longer.

The timber cruiser was also responsible for property lines being clear whether a stumpage purchase or Company land. With deed descriptions often vague or non-existent, timber trespass was a continuing problem. There are many stories of arguments—lawsuits and pitched battles—over boundary lines.

Company lands were another source of timber—cut as they matured—dependent upon markets and weather conditions. Comparatively, J.G. Deering & Son was not a large land owner. Over years the Company (and both Frank and Joseph Deering personally) acquired scattered woodlots—some with choice white pine timber. Asa Cunningham had urged purchase of timber lands in the 1930's to better insure a mill supply. Joseph Deering considered, then declined several large holdings in the Saco Valley—a mixed blessing—as it turned out. The 1938 hurricane caused a market glut and then a strong demand for wood products of all kinds developed in the 1940-45 war years. Deering Company lands totaled approximately 3500 acres in 1956 as timber operations wound down. Most of this land had been sold by 1960.

The third source of timber—logs cut by landowners themselves and landed on the river—gave Deering access to a great deal of timber scattered through the Saco Valley. Even after the river drive ended, Deering's trucks would pick up logs throughout the area. Joseph Deering said:

*With us it was very different than it was in the east [*of Maine*], because most of our operations we'd buy a hundred thousand or five hundred thousand [*board feet of*] logs from a farmer. Up the whole river valley were these farms you see, because this river was settled very early. We'd buy a few logs here, and they would haul them themselves, and this would give them something to do in the winter. They would haul them and pile them on the river bank, and then they would be scaled and put in the water and be driven down to Saco. But there might be fifty or sixty groups contributing to this, on the way, coming [*to the mill in*] Saco—[*Biddeford*]. We came from as far—the farthest I remember is North Conway, and across from the town are Moat Mountains, and we cut a lot of spruce on the Moat Mountains, and brought it down here. (1402.020-021c)

The mention of timber from the Moat Mountains raises the question of where the Saco drive actually began. It appears that in Joseph Deering's time (since 1920), logs to be driven came primarily from town's straddling the New Hampshire—Maine border and from other Maine towns throughout the Saco Valley. Logs were hauled by sled, scoot or truck to landings on the Saco or its tributaries. In the 19th century, before any of our informants were on the river, logs were also driven from the upper headwaters beyond Conway. That is the common belief among the river drivers we talked to.

As Richard G. Wood notes in *A History of Lumbering in Maine, 1820-1861*, the Saco River "was the scene of Maine's earlier lumber industry, and after 1820, logs still came down from its upper waters in New Hampshire.

Leo Bell (far right) shows up at lunch time to scale timber in 1938. Left to right, John Stearns, Virgil Kiesman, unidentified, Howard Hines, Harold Gain, and Leo. Courtesy Betty Bell Baker, Fryeburg, Maine.

Usher of Hollis was still buying logs in Bartlett and Conway as late as 1852. Mills flourished on its banks at Saco, Buxton, and Hollis".[1]

Ralston Bennett of Lovell, Maine, who began trucking logs for Deering in 1940, landed logs on the river bank for the drive:

*I took the lumber from the woods and to the river. I would have a helper with me all the time. I paid my own helper usually. Sometimes they paid the helper on the skidways in the woods...

The first year I started out I put it [*timber*] into Fryeburg Harbor, they call it. I put in at so many different places. I put in just below the Hemlock Bridge. We took that lot out of Mid Corner, they called it. I've been as far as Dayton [*where*] I put logs in. [*We landed*] them on the banks. Sometimes we had to roll them in ourselves to get the next load in. We started in the fall [and] hauled all winter. (1423.005c)

During Bennett's time working for J.G. Deering and Son the nature of his work changed. Instead of short hauls to the river's edge to land logs for the drive, he began making long hauls of sawn lumber from portable mills to Biddeford.

*What happened on most of the sawmill runs that I made, they put the sawmills on these big lots and logged them in by tractors. We have a big lot over here to Sweden, the biggest lot that I hauled from, that was right over here on Route 93... I hauled the lumber from the one on Mt. Tom, that's in Fryeburg. We had another one on Bridgton Ridge.... (1423.006c) I hauled all the sawed lumber. The logs that I hauled in to the mills was just small lots that they bought meanwhile, or lots that they owned. But they tried to set the mills up on lots that they owned that was large lots, like this one over to Sweden. (1423.007-008c)

Bennett figured that he hauled sawn lumber from the Sweden lot for at least two years running. He remembers compensation clearly:

*I was paid by the thousand [*board feet*]. The logs was all set up by the thousand, and the lumber from the mill to Biddeford was always set up by the thousand. (1423.010c)

Leo Bell was central to Deering's upriver timber cruising operation, along with the drive boss, Asa Cunningham. Leo split his time between woods work and his farm. As fellow scaler Ken Blaney remembered it:

*Leo, he worked at that lumber business in the winters. He was a farmer by summers. He diversified things. You couldn't just work at one thing, you know. He was head man in the corn shop. He had many things he could do. He was a very diversified man. (1403.017c)

The "corn shop" was a canning enterprise in Fryeburg that

On the left is Lawrence Cousins of Standish felling a white pine in the Limington or Standish area in the late 1920's. Note the wedge used to keep the cross-cut saw free. Courtesy Mrs. Jean Cousins, Standish, Maine.

canned the significant corn harvest in the Fryeburg area. Leo himself looked at his work this way, after being asked whether he spent more time on the river drive or in the woods:

> *Oh, I put in more time in the woods. Well, I had charge of a mill for Deering. Oh it was [in] different places. Had some in Lovell, Sweden, had a mill in Standish, I can't recollect them all. Portable mills, yes. Years ago, way back, they were steam. But they used to use a big heavy motor. Had to be oil. At one time we had two mills going. Erland Day run the mill and then the logging. Because I—my job was to buy the lots, buy the timber. Went everywhere to buy timber. Sometimes, well you had men over in the town of Waldoboro [*in the 1950's*], there they are for a spell, a year or more I guess. Sometimes folks had a lot they wanted to sell they'd contact Deering in his office, and he'd sic me onto them, then. Well if I didn't have anything that I could work on, I'd hunt around and find somebody that had a lot they wanted to sell. Used to do it that way lots of times. (1404.007-008c)

While Leo constantly scoured the area to purchase timber, he also supervised choppers and scaled cut timber:

> *I would be in the woods with a crew cutting. I surveyed, scaled the logs for quite a few years here. I used to have to go around from one place to another, wherever they were landing them, you know. Keep the logs scaled up. Then there was a few years there that I had charge of the crews in the woods. They might have a dozen crews around in different places. I had to keep their time and pay them every week. That would last along into April, probably. Then it would be getting ready for the drive... I would work, as I say, on the river until about the time that they got cleaned over Hiram Falls. Then I had my farm to work on. (1404.037c)

Operating

Leo Bell, along with Babe Day and other foremen, had charge of operating the dozens of woodlots that

Bucking up white pine into log lengths in the Fryeburg area circa 1940. Courtesy J.G. Deering & Son.

crews worked each year. The term "operating" is a traditional one describing the cutting and hauling of timber to a landing. Those doing the work are called "operators" or "jobbers". Logs were landed on a stream, river or pond bank—or on the ice itself. Contracts to cut and haul might also include the "rolling in" of logs come spring. As we have learned from Ralston Bennett, as truck transport improved and the drive was discontinued, logs were either hauled to a portable mill "setting" nearby or to a "yard" such as a skidway or high bank from which logs could be rolled by hand on to the truck bed.

Transporting timber was the "teamster's" job—at first by drags or sleds powered by oxen and horses—then, from the 1930s into the 1950s crawler and wheeled tractors replaced the horses' pulling scoots or single and double sleds. The "drags" were of local hardwoods fashioned by farmers, teamsters and blacksmiths to suit personal designs. Scoots and sleds were loaded by hand generally at the stump. For particularly inaccessible logs, teamsters used one or two animals to "twitch" (drag) a log with a slip chain to an open area. Efficient use of cant dog and skid pole made a difficult job easier, but it was all hard manual labor. Snow and frozen ground provided most effective hauling conditions. Long distances, steep topography, and rocky ground were limiting factors for a successful operator. The degree of difficulty labeled the job as a good (logging) "chance" or a tough one. J.G. Deering & Son might "let the job" to one or more woodsmen to cut and yard by the MBF [*thousand board feet*] the stumpage on a given timber lot. Or the Company might hire its own crews and equipment by the hour or day.

Deering's sleds, wagons, tractors, batteaus and other equipment were painted a bright yellow. Delivery trucks in Biddeford were slightly more elegant—a combination of yellow cab and chassis, red hood and black fenders—with red lettering. Joseph Deering was an inventive man. He took pride in revamping machines and equipment and in fabricating new designs. He built in the 1940's a rugged steel arch with two large rubber tired wheels approximately four feet in diameter. This was towed behind a tractor, and a wire cable winch lifted butt ends tree-length off the ground. It was an early fore-runner of today's four-wheeled

Lawrence Cousins of Standish is the teamster on his way to a landing. This photo was taken in the late 1920's in the Limington-Standish area of Maine. Courtesy of Mrs. Jean Cousins, Standish, Maine.

Rolling logs from a skidway onto a sled, Fryeburg, circa 1940. Courtesy J.G. Deering & Son.

Erland "Babe" Day (1911-1978) was a Deering woods and portable mill foreman. Deering used both horses and tractors by 1940. Courtesy of J.G. Deering & Son.

skidder. Company crews used it with intermittent success until logging operations phased out in the late 1950's.

Scaling

Although Leo Bell worked at a variety of jobs, he was one of Deering's most valued supervisors in the woods. Buying timber and keeping a dozen woods crews busy required skill and knowledge. Scaling timber—identifying how many board feet are in a sawlog—required special expertise and fairness.

Scaling or surveying, an older term, is the fine art of measuring a wood product accurately to the mutual benefit of all parties concerned: owner—operator—buyer. Log scaling and firewood measurement remain today without technological change. While pulp chips and most pulp wood are now measured in tons (not cords) saw logs are still measured manually one by one—for their net usable board foot content—a board foot being the traditional unit one inch thick by twelve inches wide by twelve inches long. And firewood still carries its time tested formula of a stack (four feet wide by eight feet long by four feet high) containing 128 cubic feet. No two saw logs are exactly alike in their characteristics. It takes consistent judgment with a fair and experienced eye to quickly assess these characteristics and to gauge deductions, if any, from the log rule in use.

Early, on the Saco River, logs destined for the drive were given river marks to distinguish ownership. Each mill or operator had his particular brand. It took so many "chops" with an axe to inscribe this mark through the bark into the wood—about half way along the length of a log. The Deering brand X◊X required twenty-four chops, give or take, applied in the woods or at the landing by the scaler or chopper. This method—laborious but effective—provided easy identification later in the river. For example, attempts to mark ends of a log became quickly obscured by the pounding on rocks and against one another on the voyage downstream. When branded, log owners could run their "sticks" together with others in a joint drive and separate again at strategic sorting gaps or sluiceways. Sometime about 1930 Saco River marks ceased to be used. Diamond Match Co. was no longer driving—relying on trucking to the Biddeford mill, portable mill production, and purchased sawn lumber. J.G. Deering & Son was then alone on the River (except for some limited use by Androscoggin Pulp Co. of Steep Falls) until the last drive in 1943.

But the scaler—in the woods or at the landings—continued to record the content of each log, stack of cordwood or other forest product. Each scaler had his own

"Swing H" and "Cross Diamond Cross" were the Diamond and Deering log marks (respectively). Slabs containing these marks were sawn from logs salvaged off the Saco River bottom in the vicinity of the main booms between Biddeford and Saco. Courtesy Rudolph Danis, Saco, Maine

system for identifying what he had previously measured so as to avoid duplication. Crayons or paint were often used. On occasion, marking axes or hammers (with a brand or logo) were used to stamp ends of logs or piles to identify the owner.

Human nature as it is, a scaler would occasionally tangle with an ingenious woodsman bent on stretching the quantity of work performed. Legion are the stories of ruses and strategies: Thin wafers sawed from the marked top end of a log previously scaled; re-arranged piles of pulp or firewood after the scaler had measured up; solidly piled wood supplemented with hidden short or undersized sticks—all clever tricks to keep a scaler on his toes. As Deering notes, the scaler's judgement was crucial to the financial success of the operation:

*There were men who made it their profession. Most of them were farmers, and a scaler had to be licensed. I don't know whether it was by the town or by whom. He had to be a rather honorable sort of person because it wouldn't take very much monkey business to cost—to be very expensive for the log owners.... We used to lose, some years we'd lose fifteen or twenty percent of our log drive in damage. You never could put your finger on exactly what caused it. (1402.049c)

Ken Blaney, Asa Cunningham's nephew, scaled for Deering from 1933 to 1944. Deering thought of Blaney as a second woods foreman of sorts, with Leo Bell of Fryeburg as the head scaler, woods foreman, and a river driver as far south as Hiram Dam. Ken Blaney recalls:

*My job was a scaler. I was lame, just like I am now. It was a job I could do. I was supposed to have been an office man. Never liked it. I didn't do it until I had to, until the last thing. I came in '34 and I went to work up in Lovell, for Leo Bell. At that time I wasn't a scaler.

I worked for Leo Bell then. That was the spring of '34. . . I can remember Leo was an awful worker. He was working for Deering. He was a woods foreman, he looked after the job, and scaled, two together. (1403.008c)

But Blaney's employment with Deering had diversity as well, particularly in the Depression years:

*Mine [job] was because the fact that you'd have to drive truck, you'd have to do anything. I hauled lumber from Center Fryeburg to Biddeford, sixty miles, and this was Merle Cunningham's truck, two trips a day, and I boarded right over across here right near where Carroll Perkins lives here. It would take me twelve hours, and I had the big sum of three dollars a day. The other fellow couldn't make two trips, so he took him off. But I had to get up to go at six o'clock, come back at six at night, with an old Dodge ton-and-a-half truck that would haul about two or three thousand [*board feet*] of hemlock. Up there where they loaded it—this will tickle you. They gave a fellow fifty cents. Now this lumber was green lumber, hemlock. He had to load it off the ground, on the truck. Now you stop to think, fifty cents a load. When I come over I got fifteen dollars a week, until I got to be a scaler. Now that meant that you had to work the five days. A scaler would get eighteen dollars a week, but a scaler on that job would work eight or nine hours. Then they would go home and do the bookwork, and turn it in to Deering, which would sometimes run in to ten or twelve hours. You never asked any questions, at that time, of what you had. (1403.017-018c)

When asked what prompted him to move from trucking to scaling, Blaney said it wasn't that cut and dried:

*Well no, there was, all the same thing. Now you realize coming up through the Depression, that [*I couldn't say*] I can do just this, now [*but*] I can't do this. No picky. College men—when I was up to Grindstone, college men couldn't get a job for seventy-five cents a day. Unbelievable, but they didn't have the experience in the woods when you could hire a good woodsman. I've seen that. They called

TABLE ONE

NORTHEASTERN TIMBER SALVAGE ADMINISTRATION

VOLUME OF A 16-FOOT LOG BY VARIOUS RULES
(1939–40 Following 1938 Hurricane)

Dia. Inch	Int'l 1/4" Saw Kerf	Scribners Dec. C	Saco River	Holland or Maine	Blodgett Rule	Humphrey or Vt
6	20	20	26	20	26	24
8	40	30	49	44	43	43
9		40				
10	65	60	75	68	66	66
11		70				
12	95	80	108	105	92	96
13		100				
14	135	110	147	142	123	130
15	140					
16	180	160	192	179	157	170
17		180				
18	230	210	249	232	197	217
19		240				
20	290	280	280	302	240	267
21		300				
22	355	330		363	287	320
23		380				
24	425	400		439	339	384
25		460				
26	500	500		507	397	
27		550				
28	585	580		614	457	
29		610				
30	675	660		706	514	

Source: Deering Papers, Dyer Library—York Institute.

this a Depression. I was fortunate to have my uncle and my father who would give me a job at that time, and Mr. Deering. (1403.018c)

Although Ken did all of these jobs as needed, he learned more about scaling as he went along:

*I couldn't have known too much [*about scaling when I first started here*], but I did know some. I graduated from high school in 1929. And then '30, boy, it was really Depression. I worked with my father, and we cut a lot for Jordan Lumber Company in Old Town. And I scaled it. They showed me how to scale and my father did. So that was my first start of it. My father used to be in business with Cunningham. They were brother-in-laws, and they worked together a lot of the time until he came over and worked for Deering. A.C. Cunningham came down from Canada, right down near where we used to live. He married my father's sister. He was from New Brunswick too. He came to work in the woods. My father's family were farmers. My father came to Maine mostly for the education. See my mother was a school teacher. She graduated from Fredericton Normal School, and one of the first classes of Shaw Business College, which was part of the school I went to years later. I went to Old Town High School, 1929. And then I worked a year, and then I went two winters to Husson College, and in those days the tuition to Husson College was $18.00 a month. But you worked for it. (1403.008-009c)

As Blaney worked for Deering in the 1930's, he became one of five scalers who traveled the woodlots, landings, and ponds regularly:

*Scaling is more a matter of judgement and honesty. It isn't really any great thing in any other way. Now Deering had four scalers: He had Leo Bell, real good, honest; Ed Smith from Denmark was an old fellow that used to scale for him; Simeon Guptill from Denmark, and then myself, and there was another one that used to work for Deering down this way [*Herman Stiles*]. He marked for Deering in the big mill... There was five of us. (1403.010c)

```
ORIGINAL
                                            Date  June 29, 1940

Surveyed from   Merton Eastman

        To J. G. DEERING & SON, Lumber Manufacturers, Biddeford, Me.

1316  White Pine    Logs to make  111,615   Feet  @ $10    $ 1116.15
_____Spruce          "   "   "   _____   "   "    $_____
_____Hemlock         "   "   "   _____   "   "    $_____
_____H. and N. Pine  "   "   "   _____   "   "    $_____

Town and Stream where turned in _____
Rule used _____

                                            Surveyor  Leo Bell
```

Log Scale Receipt

Leo Bell got right to the point on scaling when I asked him, "What was there to scaling, just how complicated was it?"

*Well, you've got to know lumber. If you've got a crooked log, it isn't worth as much as a straight one. Oh well, you can estimate. Not too much. I never felt right in estimating too much. Once in awhile there's a log that you can't get at with a rule, when they've dumped a pile of them over the river you know, over the bank. You've got to get every log. It's not fair to Deering and not fair to the man he's buying from either.

You've seen log rules, haven't you? I've got a Bangor Rule. I never used it. I don't think it's ever been used on this river. They had what they called the Saco River Rule. A log sixteen feet long and twelve inches through on the top end—the Saco River Rule gave it 108 feet, board measure, board feet. They changed that for a rule that gave, for logs sixteen foot long and a foot through, I think it was 95 feet. They used that for as long as I worked for them. (1404.038c)

Well, Ken Blaney made [a Saco River Rule] for me. I think I sold it. I knew I'd never use it. I think a man gave me $15.00 for it. Ken made it for me. It used to be [used on this river.] I can't think of what they called that rule, after the Saco River Rule. International or something like that, I wouldn't say. The Bangor Rule doesn't give so much as the rule we were using on the Saco River here. (1404.039c)

River driver Bob Littlefield remembers the Saco River Rule, and also the way it was used:

*Yes. I have a Saco River Rule, they called it, it's the one that Walter Seavey used to have. There you are, this is an old rule. I think perhaps Walter might have made that himself. Eight sided, with brass heads at various points.

Now you see this one here, that's a sixteen foot log, you see it's larger than the other brads, that is the size it would have to be to have a hundred feet in it.

It was used this way, what they would do, they would measure across the end of the log like this. And different sides here have different lengths. [*These brads*] they will tell you the feet in between. [*Points to brad I have my thumb on*]. That would be

Saco River Log Rule — 36" long, 1 1/16" diameter, octagonal shape, brass end plates and brads, steel point, stamped at one end: F. C. Deering. His dates are 1866-1939. Maker unknown. Courtesy J.G. Deering & Son.

a hundred [and eight] feet for a sixteen [foot, 12" inch top log]. (1425.011-012c)

We know already that Ken Blaney and perhaps Walt Seavey made their own Saco River Rules. Bob Littlefield concurred that they were generally made here in the area.

*Yes. He [a scaler] might not make it himself, there was always one or two people around who were handy making something. They would make two or three, it was something they could sell occasionally. That is the Saco River Rule. There were several other kinds, there was one they called the Bangor Rule, I remember. There's another one that is actually a caliper. (1425.012c)

Those interviewed agreed that there was a shift from the Saco River Rule to another rule, most likely the International Rule. It seems that the Saco River Rule was more "generous" in the board footage it gave. It would have been in Deering's interest, for a number of reasons, to shift to the International Rule. The scale was probably more accurate, and it was a more widely used scaling rule.

The scalers had a crucial role and it paid to keep trusted men on like Leo Bell:

*Yes, you got a little more money. One year, I got three dollars a day, scaling. I had ten different places to go to, every day. Then my wages for scaling got up to five dollars a day. Afterwards I had a mill up in Lovell that I was kind of looking after, I think I got seven dollars that time, up there.

The year that I got three dollars a day for scaling, choppers in the woods got a dollar and a half. [River drivers] got a little more, I think that same year that I got three dollars a day for scaling through the winter, I think they paid me three dollars on the drive, until we got over Hiram. I think the other fellows got a dollar seventy-five. The year that I drove for the [Diamond] Match Company, I got two dollars and a quarter, and my board. And I remember it was just a hundred days on the drive. (1404.039-040c)

As Ken Blaney and Leo Bell describe it, most of scaling took place at woodlots or at the landings. Bob Littlefield remembers scaling at Hiram Falls, perhaps to identify the value of logs damaged at Hiram:

*Sometimes, they scaled during the winter, as it was cut. Now, there would be times that for some reason it didn't get scaled and they would have to do it in the spring. I remember going down to Hiram Falls—it was on the west side of the river—once in the spring of the year. I was trying to think who surveyed the logs. Sim Guptill I think, and I went down to give the lengths for him. We had a bamboo pole. (1425.011c)

Roy Smith remembers logs being scaled in Cold River, a most unusual spot for scaling:

*They put some in up on the banks of the Cold River, they put pine in up there. Leo used to scale up there, and Carl Brown over here at Lovell used to scale. Put them by a raft, and they would have two men on the raft, besides the scaler. That tender would measure them for length, right in the river as they went by the raft. (1424.007c)

Roy Smith said that was the only scale that he knew of for that lot, that is, the only time the logs were scaled. It appears to be complicated business, keeping track of where timber was being cut, getting it to a landing, and eventually driving it to Biddeford. Keeping track of the timber going into the river was in the hands of Leo Bell, Asa Cunningham, Ken Blaney and others. It made sense that Cunningham and Bell were in charge, since they procured the lumber to begin with and had both worked for Deering a long time. Leo Bell chuckled when I asked him about what it was like to work for Asa Cunningham:

*Oh, he was all right. He used to build skidways in the woods to pile logs onto, and they had it built up so that the logs would roll onto a truck to haul them to the river. Of course some logs was landed right on the river bank. But some of them had to be trucked some distance. And picking out a site in the woods on a sloping ground where you could have a skidway, I remember going with him and I'd say it looks to me as though a skidway would go right in there pretty handy. Well, he might think it ought to be somewhere's else. Well, if he thought it should be there, then so didn't I. Well, if it didn't work out all right he'd shift it. And have me shift it. [Laughs.] If you built it where I wanted to in the first place, he never would have liked it. He had his own ideas, and they were good too, but they weren't always good. Oh, some folks didn't like A.C. to work for but I didn't have any trouble with him. (1404.006-007c)

A.C. Cunningham

While Leo Bell did much of the timber cruising and scaling for J.G. Deering & Son, it is clear that Asa Cunningham was in charge. Joseph Deering relied on Cunningham for both his timber cutting and river driving operations from 1926 through the last drive in 1943. Joe Deering remembered the following incident took place during one of his infrequent visits to the river:

*Cunningham, I came downriver with him one night and the whole crew of them—must have been twenty-five or thirty of these birds—had gotten drunk. They had a barrel of cider, hardened, and then they would freeze it. They would preserve the rest of it, and once or twice a summer they would have a drunk. It would be a good one. They were picking on the cookee [that night], and Cunningham, who must have been sixty years old at the time, just

White Pine on the Saco River

TABLE TWO
J.G. Deering & Son - Biddeford, Maine -
Log Quantities and Costs

	"Upcountry" "Upriver" Logs (Note 4)	Total Avg Stumpage & Operating Costs Delvd. to River or Stream [Totals of these two columns equal costs delvd. Biddeford Mill]	Total Driving Costs Alone F.C. Deering's Notebook	Biddeford-Saco "Winter", "Landing"-"Flats" or "Sawyer Boom" Logs (Notes 4&5)	Total Avg Stumpage & Operating Costs Delvd. to River Bank or Boom Biddeford Mill
(Note 3)					
1898	2140 MBF	$ 7.56 MBF			
1899	(No records)				
1900	4110	8.40		410 MBF	$ 9.21 MBF
1901	4752	8.36		1148	9.10
1902	4667	8.58		728	9.42
1903	4953	8.34	.29 MBF	864	9.57
1904	6522	9.09	.46	569	10.42
1905	6529	9.92	.41	811	10.61
1906	7224	10.36	.31	491	12.11
1907	6276	11.49	.44	726	13.95
1908	5895	11.79	.47	191	14.00
1909	5122	11.31	.75	335	14.00
1910	6699	11.92	.46	208	11.26
1911	6905	12.21	.80	143	13.04
1912	4568	11.23	.62	554	13.32
1913	3502	11.95	1.25	4	14.50
1914	3883	12.24	.54	262	12.00
1915	3160	11.68	.60	397	14.13
1916	5770	12.44	.47	84	14.75
1917	5432	12.56	.69	42	14.75
1918	3943	15.44	1.00	8	14.13
1919	3129	18.49	1.55	51	20.00
1920	3382	18.62	1.46	5	20.00
1921	3206	21.62	2.04		
1922	3097	17.78	1.23	93	17.78
1923	3222	20.90	2.25	2	
1924	3070	23.45	3.00	42	
1925	2371	19.92	2.73	2	
1926	3531	20.18	1.00	4	
1927	4585	18.81	1.16	1	
1928	2903	17.38	1.04	472	17.86
1929	4697	15.36	1.31	9	17.00
1930	4797	15.00	.99 MBF	483	15.58
1931	4162	11.83		625	14.04
1932	3914	9.66			
1933	2839	9.53		619	10.03
1934	3427 "Operated" Logs	11.21		538 "Purchased" Logs	9.31
1935	2577 (Note 6)	12.28		1842 (Note 6)	10.03
1936	2936	10.55		1410	10.00
1937	3926	11.70		1123	10.17
1938	3554	12.30		771	10.13
1939	2949	13.74		943	11.74
1940	6519	11.31		1028	10.41
1941	2865	14.38		1702	11.95
1942	4675	16.82		352	17.50
1943 Last Drive	5400	20.75		1192	21.95
1944 (Note 7)	2465	28.55 MBF		699	24.08 MBF
	196250 MBF TOTAL			21983 MBF TOTAL	

NOTES
1. Words in "quotes" as taken from company records.
2. Diamond Match Co. is referenced twice: 1928–7395 MBF (in a manner to indicate Deering took drive for them).1934–508 MBF both amounts in addition to J.G.D. & Son totals.
3. No records located for earlier than 1898.We presume J.G.D. & Son drove Saco River alone—or with others—from 1866 founding date.
4. Heaviest to white pine—up to 30-40% hemlock (particularly early years) lesser quantities of spruce & norway pine.
5. Local logs allowed mill to begin sawing early spring/summer before main body of drive arrived. Variation in annual totals unexplained.
6. Records use new terms. We believe "operated" logs were driven—"purchased" logs were probably locally delivered directly to Biddeford Boom.
7. 1943 was last full scale drive. However these records for 1944 may indicate a partial drive—perhaps from below Union Falls—including other tributaries—into the last stretch of the Saco above main boom. From then on to 1948, the last year the Biddeford mill sawed, logs were delivered by truck.

took his coat off and sailed right in and flattened them all down. And I said to him, "Cunningham, how do you dare do a thing like that?"

He said, "Mr. Deering, drunks can't fight." The woods boss and the river boss had to be physically predominant. No fooling about it, these were rough people. (1402.19-20c)

There is no doubt that the men who worked on the river were a crusty bunch, to say the least. But there are two things worth noting here. First off, much has been made of the stereotype of the drunken woodsman. Robert E. Pike's *Tall Trees and Tough Men* has its moments in describing in wonderful detail New England loggers and river men, but it also gives credence to a rather unfair characterization of a woodsman.

> In a rough and tumble fight, the riverman was probably the best man with his hands in the world, and the riverman loved to fight.... When the riverman hit town—especially at the end of a drive, with the logs in the booms and his pay in his pocket, with the memory of months of hardships, of backbreaking work and narrow escapes from death behind him—he figured it was his duty to rip the town apart. It might take from three days to three weeks for one great and glorious bust, while Godfearing citizens locked their doors and the local police looked the other way, before the riverman had thrown away all his money, and broke, sore, and sober, headed back to the tall timber.[2]

Pike relates to us a commonly held belief. It was a hard life, especially for those, for example, in the Penobscot River lumbering operations, where they did go into the woods or worked on drives for weeks and months on end. Work on the Saco River was quite different. Unlike much of northern New England, the Saco River Valley was settled early and was quite agricultural across its 120-mile route through New Hampshire and Maine to the ocean. As a result, with the exception of a few overnights at boarding houses on the drive, by and large the woodsmen on the Saco went home every night. There were few weeks-on-end away from home. Nearby woods or river work was an integral part of a Saco Valley family's daily living.

Later in the interview with Joseph Deering, at his summer home in Biddeford Pool, he reminisced in more general terms about his long-term, trusted woods and river boss:

> *A.C. Cunningham, everybody called him Asa. He was the foreman. A wonderful [*woodsman*], he must have been born in the woods. He lived up north of Bangor. He worked years for me. My guess would be twenty or more. He moved into this area, lived here all the time. He bought my timber and cut it and then drove it down the river. (1402.041c)
>
> *Tom McDonough said, "I have this man here who used to work for me." [*McDonough*] used to be the foreman of the Stearns Lumber Company, and they had a sawmill somewhere around Bangor. "He's out of work right now." Unsolicited, this man came into my life and he was one of the greatest factors I had in my life, he could do anything. He was just a marvelous. . . . (1402.055c)

The Drive's End

After the arthritis in Cunningham's legs crippled him in the 1940s, Deering found that he literally could not replace him. Even with a year or two reprieve assisted by Leo Bell, Carroll Perkins and others driving Cunningham around, serving, as Leo put it, as "Cunningham's legs," Asa's health was one of the principal factors in Deering's decision to stop river driving on the Saco.

It is important to realize just how much timberland could be efficiently reached through the dense network of ponds and small streams that eventually lead to the Saco River. Early efforts by lumbermen in the first two decades of the 19th century resulted in the Moose Brook Canal (1807), Hancock Brook Canal (1811), and the Fryeburg Canal (1815-1836).[3] Deering took advantage of this geography. From Kezar Lake, Cold River and Lovewell Pond in the Fryeburg area to Moose Pond in Bridgton to the Great Ossipee which flows into the Saco at Cornish and the Little Ossipee entering at Limington, the effective "reach" of the Saco River into the hinterland for timber covered much of York, Cumberland, and southern Oxford counties. The end of the river drive was to change the Deering lumber operation, hastening the switch from producing lumber for the wholesale market to retailing lumber products for the general public.

As Joseph Deering himself wrote shortly after the decision to end the drive permanently:

> The hurricane of 1938; the black market of World War II, and the forest fires of 1947 so changed the picture that log driving had to be given up. It was a sad decision because I have been always conscious of the old adage: "From shirt sleeves to shirt sleeves in three generations." My grandfather started out with nothing, and the breaking of a long established method could look like the failure of the individual involved.[3]

It was clearly a decision Joseph Deering did not take lightly. As he wrote in September 1943 to Herbert Locke, the lawyer he used for many years to represent him in his dealings with Cumberland County Power and Light Company and Central Maine Power, "Our driving difficulties this spring make it exceedingly doubtful if we shall have a drive next spring. We are attempting to buy our logs around Biddeford because we just couldn't get the drivers on our last spring's drive and if the War continues it will be totally impossible next year."[4]

Joe Deering still hadn't entirely given up hope in 1947. In four letters, written in May, August, and September of 1947 and in February of 1948, Deering queries both Fred Gordon and William F. Wyman, then president of

Central Maine Power, as to provisions being made for the passage of logs at the dam being built at Union Falls.[5] In May, Deering indicated a desire to have "a moderate drive coming out of the Little Ossipee and down Bonny Eagle Pond next spring." In August, Deering wrote to Wyman saying, "Because of wartime controls we were obliged to discontinue driving logs on the Saco River. We had hoped to start last year but labor conditions prevented it and, consequently, we are looking forward to a short drive on the Saco in the spring of '48." Finally, in February of 1948, Deering wrote, "As a result of the fires we have had to change our plans and it is now apparent that in cutting the timber necessitated by the fire we shall not have a drive next spring."[6]

As we close this chapter on timber cruising and operating woodlots it cannot be over-emphasized how crucial the river was to getting the timber out. Circumstances beyond Deering's control—a general labor shortage in the war years, constant confrontations with power companies on water levels and sluiceways, the 1947 fires, the improvements of roads and trucks, the efficiency of the portable sawmill—all conspired to bring an end to river driving. On top of all this Deering found himself without a replacement for A.C. Cunningham, the master driver and woods boss of twenty years. As we shall see in the following chapter on the drive, Cunningham was able to keep a very complex operation underway. Charlie Foran recalls how much Deering wanted him to take over the drive:

> *Joe wanted me to go up and take a log drive down the river. In fact he begged me to do it. I said, "Joe, I can't do it, I don't know anything about it. Those fellows will drown me. I remember one year, Joe, when Cunningham let the logs go on the high water, they went in on everybody's land. I remember just as today, cost you three dollars a thousand to get those back into the river. And you never did let up on him. God help little poor Charlie, what would you say to him?" I didn't do it. He wanted me to, but I didn't do it. (1407.025c)

Deering's response was characteristically unrelenting—"If I'm willing to take a chance Charlie, you ought to." (1407.074c) Charlie continued the work he knew—manufacturing lumber at the Deering mill, yet even at that time the company would continue the manufacture of lumber for less than ten years. Charlie explains the effect on production of no "up-river" drive from Lovell and Sweden.

> *Deering owned lumber, at that time, up around Lovell, Sweden, and of course they used to bring the drive down the river. I forgot what year it was, Mr. Cunningham, who used to operate the woods business and bring the drive down the river, he got used up and he retired. Deering owned a lot of land around here, he had bought a lot. Deering put me taking his place, in the fall after the mill shut down. He said, "Charlie I'm going to put you in Mr. Cunningham's place, cutting the timber." I says, "Joe, I can't bring it down the river, I never worked on the drive." Anyway, he said that we'll cut around here this winter. We did. We had several operations in Westbrook and all around. I had enough timber cut that we put on the banks of the river in different places to run the mill. In the spring we rolled the logs into the river and we ran the mill. We used to saw around five million, but we didn't get five million cut around here. We didn't have time. So that year we didn't cut quite so much. That went all right. (1407.007-008c)

It must have been a trying period for Joseph Deering. The hurricane of 1938 wreaked havoc by blowing down Maine and New Hampshire timber and glutting the market. His father died in 1939, and his trusted drive boss was less and less able to get around with each passing year with the arthritis in his legs slowly crippling him. And then World War II, with its effects on manufacturing, created a real demand for lumber for wooden shipping crates. On top of all that, Deering commissioned a study by the James W. Sewall Company, published in 1943,[7] which told Deering this:

> *I had in the '30's the fellow, his son is president of the Maine Senate now, Sewall, Jim Sewall became a friend of mine. I had him make a survey of the timber in the Saco Valley. He came up with the fact that there was a billion feet of merchantable timber in the Saco Valley. I just felt so pleased with myself because nobody knew it.
>
> The hurricane came in 1938 and tipped over all the timber in New Hampshire. It didn't leak over much through the mountain range. It went up through New Hampshire. But then all the damn portable mills in Christendom moved over into the Saco Valley, almost over night. I drove for ten years more, fifteen years more, then had to give up log driving. (1402.008-009c)

Deering's recollection that he drove ten or fifteen more years after the 1938 hurricane does not reflect reality, especially since the last drive occurred in 1943. What his recollection does reflect is his unwillingness to give up on river driving. His correspondence with Wyman at Central Maine Power in 1948 shows he still had not given up on the idea entirely.

Giving up log driving effectively shut off that wonderful, continuous source of timber by water. The vast territory Deering was able to operate by using water transportation on the Saco and its adjoining lakes, ponds and streams would be altered. Although the increasing use of trucks and portable mills provided raw material, the scale of lumber manufacturing in Biddeford was not to be matched again. A legacy of nearly 300 years—the lumberman's continuous use of water transportation to his mill—was coming to a close.

Page from record book of the Proprietors of Saco Boom incorporated 1805. Stock of this corporation was held by the Deering family from the early 1900's until its liquidation in 1959. Courtesy J.G. Deering & Son.

CHAPTER THREE

The Saco River Drive

According to David C. Smith, "the two most important items in any river drive were first, control and conservation of the water, and second, a crew of men who would do the actual work".[1] We will learn about the Saco River with its natural difficulties of meanders, bogs, and falls, its wide sandy plains, and its man-made obstacles such as power dams. We will hear about driving the Saco from the men who did the work. As Emerson Baker has demonstrated, it is a story that begins with the establishment of sawmills in the seventeenth century and eventually the first log drive down the Saco River from Fryeburg in the spring of 1772. The incorporation in 1805 of the "Proprietors of Saco Boom" marked the continuing development of river driving on the Saco. The river drivers of the twentieth century complete the story.

The Saco River—Background

This book is concerned chiefly with twentieth century river driving operations of the J.G. Deering & Son, its subsidiary, Saco River Driving Corporation, and the Diamond Match Company, particularly after 1920 when the oral history record is available. But it is worth noting some very much earlier history, already touched upon by Emerson Baker in an earlier chapter.

The Saco Boom, one of the first attempts to establish a log impoundment on the river, was incorporated under the Massachusetts Legislature in 1805, pre-dating Maine charters on the Penobscot by ten years.[5] Foxwell Cutts, Seth Spring & Sons, Abner Sawyer, Daniel Cole, Thomas Cutts, Samuel Dennet, David Bryant, James Gray, Noah Hooper, Nathaniel Cole, James Carlisle, and Moses Bradbury, "together with such other persons as now have or shall hereafter become proprietors in the said Boom, be and they hereby are constituted and made a corporation for laying and maintaining a boom across the Saco River, by the name of the *Proprietors of Saco Boom*."[6]

Over the years from 1805 to 1866, we know there were hundreds of sawmills and timbermen on the Saco. To our knowledge there is no collection of factual information on this industry during these years or into the twentieth century. The references that follow are available from various sources. However, they serve only as fragments of what we know to have been such a vital and substantial business for Saco Valley inhabitants. Later in the century, after J.G. Deering had entered the lumber business in 1866, we know that Joseph Hobson of Saco was one of the biggest operators, with eight million board feet of lumber in the boom by June 1st.[7] As David Smith notes, the drives on the Saco were usually quite early in the season. In 1878, the first drive of 1.5 million feet was in by May 1. Eleven years later, in 1889, the total of logs in the Saco drives totalled 17.5 million feet. All logs had cleared the boom by August 13, and according to local newspaper reports, this was the largest cut on the river in some time.

Smith also notes that in the 1880's, "This area, around Hiram, Fryeburg, and Bartlett, New Hampshire, continued to produce three or four million feet each year for the Saco mills, mostly spruce and hemlock". Farther south, three million board feet of pine were cut on the Little Ossipee River in 1875.[8] In the 1940's, the Deering Mill still manufactured anywhere from 3 to 7 million feet of pine, coming principally from the same region, particularly the Fryeburg area.

The Saco River Basin, or "drainage district," as Walter Wells calls it in *Water Power of Maine* (1869), encompasses about 1400 square miles. The river's length, from Mt. Washington to Biddeford is about ninety-five miles. It varies in width—from narrow and quick to wide (500 feet) and lazy. The chief deviation from a southeasterly direction is in Fryeburg, where prior to the construction of a canal across an enclosed peninsula, the river loops to the north meandering around Fryeburg.[9]

Wells goes on to note that the upper part of the Saco is situated in a cluster of the White Mountains so as to "carry off a considerable part of the surplus waters that would otherwise be discharged by the Merrimack."[10] The descent of the river is moderately steep, the current generally moderate with the greater part of the descent taking place in sudden breaks or falls at Conway, Swans Falls, Hiram Great Falls, Limington, Bonny Eagle Falls, Buxton, Bar Mills, Salmon Falls, Union Falls, and Saco Falls.

One outstanding feature is the great number of lakes and ponds in the Saco River Basin—Wells gives the total number of lakes as seventy-five, more than twice the average concentration for the whole state. Other major tributaries flowing into the Saco are the Cold River (12 miles long), Kezar River (14 miles), Kezar Outlet (3 miles), Moose Pond, Hancock Brook, Great Ossipee (33 miles) and the Little Ossipee (30 miles).

Types of Booms on the Saco River

While water levels were crucial, so were the tools of the river driver, used primarily to direct and control the flow of logs. For example, a variety of log booms—sorting, sluiceway, fin, holding, and tow booms—were used all along the Saco River. A boom "is a series of long logs chained end-to-end together."[2] As its name implies, a sorting boom enables drivers to sort logs into different groups according to their log marks. As we learned in the previous chapter, it was the chopper or the scaler who applied, in Deering's case, the "cross diamond cross" mark on the side

Fin Boom in operation. Courtesy C. Max Hilton, Rough Pulpwood Operating.

of the log with an axe. By the twentieth century, sorting on the Saco occurred in just two places—at the Steep Falls pulp mill and in Biddeford where Deering logs went to the east side of the river and Diamond Match logs went to the west. Leo Bell remembers:

> *There was a pulp company at Steep Falls.... Their stuff was all put on one side of the river, the Deering and Match stuff went right down by it. There was a line of piers right up and down the middle, up above Steep Falls. And a boom, a sorting boom up there, and anything that was going down river was kept on one side of the boom, and anything that stopped at Steep Falls went down on the other side. The river was split. (1404.004-005c)

When asked if there where any other sorting booms on the river, in addition to the Deering/Diamond boom in Biddeford, Leo responded:

> *Not at that time, not in my day. Years ago, when I was ten or twelve years old, there were two or three different companies. I know there used to be at least four river marks, "Cross Diamond Cross" and "Swing H" and then there's "Crow's Foot". Cross Diamond Cross was Deering's, and Swing H was Diamond's. In my time there were only three companies, Deering, Diamond and the pulp company at Steep Falls [*Crow's Foot*]. (1404.005c)

Ed Burrill remembers the Steep Fall sorting gap as a simple affair:

> *The sorting gap was there at Steep Falls. See downriver logs went one side, they had the river boomed out, downriver logs went one way and the pulp company's went in on the other side [*east*] of the boom.
>
> [*The pulp company logs*] were whole trees, not pulp length [*four feet*]. You could tell the difference because most all of the pulp company's was spruce and fir and poplar. They used an awful lot of poplar

for paper then. Deering's logs was bigger. I think the length of their logs was ten, twelve, fourteen, sixteen. That's the way they cut them. (1405.009c)

It appears that Ed is relaying information passed on to him. He didn't recall log marks perhaps because he didn't see them himself. His first year on the river was 1929, when the Steep Falls Mill did not drive and after the Diamond Match Company stopped driving logs. Bob Littlefield remembers the pulp mill from his first drive in 1920:

Leroy Smith (1908–) and Herbert Kimball of Fryeburg towing a boom to the outlet of Moose Pond in Bridgton in the late 1930's. Logs were driven from Moose Pond through Moose Brook to the Saco. Courtesy Leroy Smith, Lovell, Maine.

Ed Burrill describes two types of boom hitches — toggle chains through a hole in the log (left) and slip ring chains that would go around logs (right). Photo courtesy Edmund A. Choroszy, Saco, Maine.

*That was at Steep Falls. Now that pulp mill was there the first year that I worked on the log drive.... That would be four foot stuff. There would be a little of it mixed in. At that time, we never bothered to separate it. It wasn't supposed to be mixed in with our logs. It was all four foot. They drive separately, just as soon as the long stuff went through, they would come with the pulp.

There was [*a sorting boom*] at Steep Falls, because sometimes they had so much pulp mixed in with the logs, they had to sort it.

They would make what they called a work boom, a couple of good sized logs toggled together so they couldn't turn, couldn't roll under, so it would be handy to work on. Then they would have this boom fixed so that they could let the short stuff through in one space and the long logs would go in the other. They would just have to sort it as it came there. They would do it with poles, pickpoles, we called them. (1425.013c)

Although there appears to be some disagreement as to whether the pulp was four-foot or longer length, one has to surmise that the pulp probably came in varying lengths. If it were all four-foot, then there would not have been a need for the Steep Falls "crow's foot" log mark.

Sluiceway booms or sluiceways guided logs to where you wanted them to go, or, on the other hand, kept them from going where you didn't want them to go. Cribwork piers were crucial for booms that guided the logs to the sluiceways over dams at Hiram, Steep Falls, Bonny Eagle, Buxton and Bar Mills. In addition, there was a sluiceway above Limington Falls. Bob Littlefield explains:

*I am sure that anyone could find those old piers up at the upper end of Limington Falls. Sometime go up and go on the west side of the river, there's a little dirt road that goes up through there by another short set of rips, Parkers Rips, I think it was. You can see where the falls start, and right there on that side you will see these old crib piers.

What we had there was a good boom put across from those piers out toward the east side of the river. That was to keep the stuff from running full width down those falls. If it went like that it would make an endless job to get the stuff picked up again, due to the fact that the water would drop and leave them dry.

I'm not sure whether there was any pier above Hiram Falls, I don't remember of using one. Trees in there were too handy to tie to. They were all right. (1425.014c)

Clarence Brown helped build cribwork piers at Bonny Eagle:

*They had a boom there, at Bonny Eagle, I helped build some of the [*cribwork*] piers. Built them in the winter, on the ice. Cut a big hole in the ice there, so that you can build the crib up there.

White Pine on the Saco River

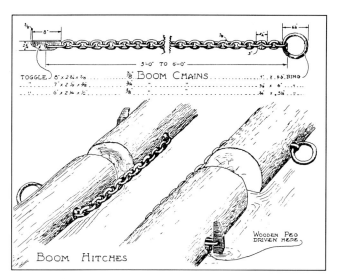

Courtesy of C. Max Hilton, Rough Pulpwood Operating.

The first logs that you put down, you've got to sound, take a measure down on that corner and that corner, so that you know, you've got to make the first logs fit the bottom. So, we put the other logs on it and keep building up and keep loading it with rocks, four feet or something at the hole, logs all the way across. At the floor, load that with rocks until sank down at water level, then you can build up again. When it went down, them first logs that you made, they'd fit that bottom and be upright. (1406.020c)

The boom was then chained to the pier: "So that the logs could go along the side of the boom. If it was on the front side, the logs would catch on it. The boom went by your pier, so that the logs would follow your boom right along. (1406.021c)

In addition to attaching booms to cribwork piers or a hardy tree, fin booms were used to hold logs from backing into dead water or simply going where they would be high and dry if the water level dropped. The fin boom used the river's current (*see page 33*). Hung at the upriver end to a cribwork pier or the shore, "the lower end of it hung free, but the action of the current on the fins forced it out far enough into the current to deflect the logs" into the downriver channel, not into dead or shallow water.[3]

A holding boom simply held logs at a certain point in the river until the crew was available to guide them into a sluiceway or to wait for water levels at the dams to be set at the correct level with splashboards. Ken Blaney recalls the major holding booms, particularly how holding booms and sluiceway booms were set up together:

> *They went down to Hiram and put a boom across above the dam. Then they let the logs down through, so many at a time, down through the sluiceway there, in Hiram. So they held it again in Steep Falls. That was two of the big holding places, there at the Steep Falls bridge. They would have about five or six million, in there. There was five or six dams down through. I used to do a lot of sluicing when I didn't scale. You'd stand out on the logs there and keep the logs separated when they went down through there. (1403.012c)

Walter Casey at Hiram Falls, sharpening his peavey against the bow of a batteau in the 1930's. Courtesy of J.G. Deering & Son.

The tow boom was used in moving large groups of logs across any of the several lakes and ponds. Logs were skidded onto the ice in the winter, and the ice would melt from under them in the spring. Since there is no current in a lake, booms would be connected together around a mass of logs and towed either by capstan raft or, in later years, a motor boat. Vast amounts of timber were boomed across Kezar Pond, Lovewell Pond, Pleasant Pond, Kezar Lake, Moose Pond, and others. Once the logs were at the entrance to the Saco River, and once the river drivers were in place to handle them, the boom would be opened. Although it seems on the surface to be a most efficient way of moving logs (and it probably was) the process was not without its problems—including headwinds that made it hard to move a large boom, and unreliable motor boats.

Ed Burrill sums up the use of chains and how the booms were made, how the holding boom worked, and the practice of sweeping coves and eddies for stray logs:

> *If they had a long stretch of river to divide, they had whole trees, for the boom. You took them and ringed them, cut a ring around them right in deep, and there was a big chain with a fitting link on it. You put that chain to that the same as you would a string and pull it into that notch and hook it to the other until you got your end [of the other log] up there.
>
> That chain, a boom chain.... If we had a place where we had to sweep an eddy, we had toggle chains. Just toggle some logs together with pins. Got your boom around your logs and sweep them out. If you had forty- fifty to a hundred sticks of them tied together, you had to have quite a chain to hold them. At the piers, where they caught the logs to hold them 'til we got the rear in there, before they cut the boom.... They'd have a boom across, and there would be a pier out there, you hook that into the pier and the other one to the shore. When you cut the boom, you went out and took the cable off from that end over there and let it swing to shore and then your logs could go.
>
> The boom would be hooked down on the back end of the pier. If you put it up front, your logs would hold it there, you couldn't swing it. When you cut it, down on the back corner, your logs would push it right ashore. (1405.016-017c)

Clarence Brown and Ed Burrill have differing opinions as to whether the boom was attached to the upriver or downriver side of the pier. Brown contends it was at the front of the pier so the logs would not catch on the pier itself. Burrill remembers cutting the boom to let logs go downriver, and it is clear that you could not cut the boom entirely free if you had the current and the weight of logs pinning the boom log against the pier. It is worth pointing out that interviewers for the book *Argyle Boom* found the same differences of opinion among several Penobscot River drivers. Although Ives *et. al.* chose to illustrate the booms on the downriver side of the piers,

Teamsters Joe LaCroix and Norman Boothby sledding logs to Lovewell Pond, Fryeburg, circa 1940. Courtesy J.G. Deering & Son.

they wrote, "we would prefer not to consider the matter settled."[4] They go on to say that perhaps the practice varied, and I would submit that on the Saco, the booms were attached one way or the other according to use. That is, if it was a holding boom that would be cut, the boom was set below the pier on the downriver side. When cut, the boom would swing free. If it was a sluiceway boom—the type of boom most familiar to Clarence Brown—then of course the drivers wanted as few areas as possible for the logs to catch on, so they set that boom in front of the pier, enabling the logs to slip right by.

Burrill also describes the tools of a river driver particularly the cant dog or peavey (the manufacturer's name) and the pickpole:

> *Long handled ones. You probably won't believe this, but Cunningham, you can ask Joe Deering, if he can remember it. One time down on Salmon Falls, on that rock, we lost twelve peaveys. Brand new ones, they just bought them. You see,

Teamster Joe LaCroix, on the first double sled, heads out onto Lovewell Pond, Fryeburg. Norman Boothby (standing) is with the second team. Pleasant Mountain is in the background in this eastward view taken about 1940. Courtesy of J.G. Deering & Son.

you get hooked onto a log and you couldn't let go. You couldn't get that dog out. They had to be made [*sharp*] by a blacksmith, so when you yanked back on it, you'd yank it out. Them weren't, and every one of them peaveys is in that river right there by Salmon Falls. You couldn't let that peavey pull you in. Right there by that rock that they used to pull us off from, them logs used to go down and go in under a ledge, and when they come out the other side they'd be all peeled, where they had twirled in around them rocks. You couldn't very well let a peavey haul you in there. Cunningham used to holler, "Save that peavey, never mind the man." (1405.017-018c)

Clarence Brown speaks to the need of sharpening the peaveys: (*see photo page 35*)

*Peavey Manufacturing Company made the ones we had. Bangor. Hubbard turned the handles for the Peaveys, in the mill here in East Limington. He made the best handles that you could get anywhere in the country. Charles Chase's blacksmith shop was just a little ways from there, and they put the irons on and fixed the points. They'd lose them and break [*peaveys on a drive*]. You know how the point is, if that point is sharp, they heat it so that the corners are sharp, you can reach out and stick it into the log. You can pull that log just by sticking into the end of it. If you lose them corners, you can hit it as hard as you want to, the log, and the darn thing will jump right back out, won't hold. So you had to take them back and have those points hammered again. The pickpole is the same. (1406.014-015c)

Having summarized some of the basic uses of river driving booms and tools, we can now look at some other features of the Saco River drive.

In earlier chapters, we learned about Leo Bell and Ken Blaney's work in the woods. Ed Burrill, Bob Littlefield, Roy Smith, and Clarence Brown add to the story by telling of their first hand experiences on the drive. These men provide the story of what it was like to work on the river, particularly what the drive entailed, from April to as late as August of any given year. Ed Burrill of Limington, who now lives in Cornish, was one of the "white water men", to use Ken Blaney's phrase:

*"Well, we went up in '29 and, they drove every spring until 1943, and '43 was the last one. Of course [when] I was young, just a young kid, I went back on the rear, kind of a water boy, you know. But when I went up in '29, I was on the river about every spring. . . (1405.005c) *It was something that got in your blood, and you kind of look for it every spring. (1405.007c)

Bob Littlefield of Lovell started work on the Saco River drive for Deering in 1920, worked for four or five years, and then later in 1934 he drove again. Born in June 1903, he was seventeen years old when he began in the spring of 1920.

*I was young, with lots of vinegar, you know, and it was fun, although it was hard work, it was fun for us young fellows. I started work for [*Frank Brown, Deering drive boss*], I worked three or four years before A.C. Cunningham came. Skiff Kelso was working as foreman for the [*Diamond*] Match Company. (1425.004-006c)

Littlefield tells of the earlier days, up until the early 1930s, when Deering and Diamond drove the river together. Roy Smith drove the river with Ed Burrill, Leo Bell, Merle Cunningham, and others in 1930s and 1940s. Roy Smith boarded at various points down the river to Biddeford. He was also a "white water man," and frequently worked the batteau.

What we know about Frank Brown, Deering drive boss from 1909 until his death December 18, 1923, comes from his son, Clarence, who began driving when he was nineteen. Clarence normally tended sluiceways at West Buxton or Limington and he worked the drive itself for three years. More importantly, because his father was a drive boss, Clarence has collected much information on the drive and the men who worked on it (*see appendix three*).

From previous chapters, we learned about the cruising, operating, and scaling of Saco Valley timber. Saco Valley pine was at times cut no smaller than eight inches in diameter, chest high. The photographs of the landings on the ice at Lovewell Pond indicate the average size of much of the timber skidded onto the ice with horse drawn sleds and scoots. From there, with ice out, the logs were boomed up and pulled by boat to the Saco River.

In both the size of drainage basin and the number of men employed, the Saco River was much smaller than the

Landing area at Fryeburg Harbor on the old course of the Saco River, circa 1920-1930. Courtesy Robert Littlefield, Lovell, Maine.

massive Penobscot operations. Generally twenty or so men drove the Saco, where as many as 100-200 men were involved in the comparatively huge Penobscot drive and sorting operations at Argyle Boom which handled timber from "that great 10,000 square mile drainage basin... from Jackman to the Mattawamkeag."[11]

The Drive Begins

Asa Cunningham's journal reports in 1940 that by April 21 the driving season was one month late. "Snow has not begun to melt and logs are frozen into the landings." In addition to logs on the ice, Roy Smith remembered there were also landings on the river bank throughout the Lovell-Fryeburg area and downriver:

*Over here at North Fryeburg, on the Old Saco, they had that full. They had some down to Toll Bridge... before you get to the covered [*Hemlock*] bridge, that's where they had the main boom (Hobbs). The logs was all in that, and when they got ready to drive they cut the main boom. Of course, up where they started, there would be ice. Had to pry the logs out of ice, feed them into the river. (1424.005-006c)

Smith goes on to confirm much of the early work took place in ice and snow:

*The logs, they all put into the Harbor [*Fryeburg Harbor*]. Then they cut logs, when I was working for Diamond, we cut logs way up at the head of Cold River, peeled hemlock, we put them and drove them down to Charles Pond. Leo was there and the pond was frozen, and we took ice saws and cut the channel. The outlet of Cold River came into the pond, cut it right through to Charles River, where it went out. [*MC: You cut a channel through the ice?*] Floated the logs through it and that saved putting booms along each side. That was cold work too. (1424.006c)

Leo Bell remembers both landing logs on the ice in ponds and on the river banks.

*Start in early spring, soon as the ice was out of the river. Course there were different seasons, different times to start. Started up in Conway, [*NH*]. Years ago, before I ever drove they used to start up in Bartlett, that's north of Conway. [*MC: Would the logs be at the river banks?*] Well, up that way they would

Log landing on the Great Ossipee River. Courtesy Mrs. Sondra Pinkos, Cornish, Maine

have to be, because there's no ponds to put them into to drive them to the Saco River. Years ago, up at Kezar Pond there, they used to land logs on the ice, and then the ice went out of the pond, why then they put a boom around them and drove them down the pond to the Saco River.

[*Referring to the photograph on page 37.*] Yes, that's on Lovewell Pond. Yes, I remember that. The outlet is way down here [*to the right*] somewhere, about two miles. If they weren't on the pond they'd have to be landed on the bank of the river. [*The landings on the bank*] piled up, just like a load of logs [like on a skidway].

If there was much slope, that sloped down to the bank, you'd have to drive stakes on the river bank. Let the logs come down to that, well then when you get ready to move you've got to cut them stakes and roll them into the river. (1404.008-009c)

Leo Bell raises an interesting question as to where the drive actually "started." The lion's share of the logs, as he described it, were skidded onto the ice or landed at brooks and the river's edge. Many of the river drivers felt certain that before their time, the drive began at Bartlett and Conway, New Hampshire. No one I talked to has direct experience of the drive starting upriver from Fryeburg Harbor. Clarence Brown remembers the landings this way:

*Canal Bridge [*on New River*], the first bridge you come to going [*on Route 5*] from Fryeburg towards Lovell. Next one [*Toll Bridge on Route 5*] is over the Old River. We started those logs right up in Lovell village. Come out of [*Mill*] Pond [*on Kezar River*], there's a mill and a dam right in the center of the village. They had logs piled all along the banks of that. It's just a small stream. The only way you could get them out was on your early floodwaters. (1406.005c)

A look at Asa Cunningham's journal tells us that a great many logs were also landed on both the Little Ossipee River and the Great Ossipee River, tributaries which extend directly west into prime timber land. Cunningham started Ossipee logs *before* logs from Cold River and Kezar Lake. As a result, he simultaneously ran a drive beginning nearly halfway down the Saco (where the Ossipee joins it), as well as upriver in locations north of Fryeburg. Bob Littlefield, who drove early in the 1920s and in 1934, remembers the scattered nature of the drive's beginning. He also confirms the importance of the Ossipee River:

*[*The drive would start in*] the latter part of April, depending on the year. Here, it would start, they would go up to what we call Ring's Landing, up the Kezar River. There was usually logs up there. We got those down through into the river, and then we would go off to some other place, might be Charles Pond, perhaps up Cold River. They were around in several ponds, well, there was Cold River, Kimball Brook, of course, comes out of Kimball Pond, we had logs up there sometimes. Always some up the Ossipee River, down to Cornish. They were scattered pretty well.

Bob Littlefield also remembered in the early 1920's when Everett Boulter would come up to the Lovell area to buy stumpage for Diamond Match. He remembered that they had to identify a place to land them where they could eventually be taken to the Saco River through ponds and brooks:

*Moose Pond was a place that I think of, they would be way at the upper end of Moose Pond. We had to come through two of those bridges, sluice at Denmark, down Moose Brook, and eventually get our logs down into the Saco River.

Sid Frost of South Hiram "had charge" of the Saco River landing above Cornish Bridge on the West Baldwin side. Company records call this the "Murch Lot." 1,920,000 board feet were delivered to the landing for $4.15 operating cost and $7.85 stumpage per thousand board feet. As many as nine teams of horses with scoots and sleds yarded this white pine to the river bank from an area of more than a mile of river frontage. Photograph taken in the winter of 1934-35. Courtesy of Ken Blaney, Cornish, Maine.

What they would do, as soon as we got down through sluicing at Moose Brook, sluicing at the Lower Bridge we called it, the little one, then they would send a couple or three of us down to Denmark and we'd stay right there and sluice until they were through. (1425.006c)

The southern tip of Moose Pond is in Denmark and extends north for more than eight miles with its northern tip about two miles east of Kezar Pond. Although Moose Pond is four miles east of the Saco River in Fryeburg, all the timber landed on it was sluiced through Moose Pond Brook and entered the Saco River about five miles north of Hiram Dam, as the crow flies. The expanse of territory, the number of men deployed for sluicing, and the timing for rolling in was an extremely complicated matter:

*Up in this area, the main boom was down the [Frog] Alley Road, we call it. That's going from just below the first [Toll] bridge, there's a dirt road that goes down and comes out at East Fryeburg, down by the old covered bridge, Hemlock Bridge. We had our main boom [Hobbs] there. I've seen that filled up from there way back through to Fryeburg Harbor. What we were waiting for was for the water to go down on Brownfield Meadows. It would flood right out over the meadow in the spring. We had to get that inside [the riverbanks] and then go as soon as you can, because the water is dropping with you. (1425.009c)

Ed Burrill recalls where the main drive began and the importance of cutting the boom and freeing the logs once the water started to drop:

*Well, never but once that I know of that we was up in Conway, the furthest we went was North Fryeburg. 1929 we went up to North Conway, just above the covered bridge and the railroad bridge there, that's where we started in. But most usually it was North Fryeburg, because they cut their timber and hauled it to the bank of the river around Lovell, Sweden, like that. They had them [logs] in ponds up around Lovell you know, like Gary's Mill Pond, that was quite a big brook [Kezar River] that come in there... [and they would] drive it out on high water into the main river and have a boom across to catch it so it didn't go. And when the water started to go down they'd cut the boom and let it go. I don't know what the name of the brook was, but it run right down through and dumped into the Saco River. (1405.005c)

Map 2 on page 10 shows the area that Ed Burrill talks about. Although the drive crew had Cunningham's experience and the manpower to "turn over" landings at the appropriate time, Ken Blaney also points out just how crucial timing was in the beginning of the drive:

*Teams would bring logs in on a flowage they called it, which they brought from there down through to the Saco. It isn't a freshet, a freshet is when you have an excess, a spring flood. They didn't really have to have a flood. The normal snow would make these brooks come up enough so that they could run the logs that they put in down into the Saco. They used to have to wait before they started the drive until after... the first run-off. You couldn't put logs in then because you couldn't hold them. They would go all over everything. They'd flow the banks and they'd be out there, and when the water dropped, to get them back in would be expensive. (1403.007-008c)

Left to right, unidentified, Charles Tenney of Denmark, Frank Brown of Standish, Alonzo Foster of Steep Falls, Skiff Kelso of Bar Mills, and Hal Whitehouse of Buxton. This photo, taken prior to 1924, includes three drive bosses: Frank Brown (drive boss 1909-1923), Hal Whitehouse (1924) and Lon Foster (1925). A.C. Cunningham began in 1926. The Model T truck was used to shuttle the river crew. The photograph was taken at Lon Foster's house near the Steep Falls bridge on the Limington side. Courtesy Arnold Foster, Limington, Maine.

Picking off Hiram Dam, circa 1940. Courtesy J.G. Deering & Son.

The expense of logs escaping the riverbanks was felt keenly downriver at the J.G. Deering mill. As yard foreman Charlie Foran puts it:

*The [*first*] logs never got down here usually until sometime in May. It was supposed to be all one drive, but some logs would get [*hung up*] on the side of the river, they'd have to bring the stragglers in... It's according to how the river was. If it was extra high water they had to wait, they were supposed to wait until the water got inside its natural banks. If they didn't,they go on your land, her land, my land, everywhere. They did one year and it cost them like the dickens to get them back in the river. They usually waited until the river was back in the stream where it belonged, then they'd let them go. (1407.018-019c)

Joseph Deering was well aware that it "cost like the dickens" in more ways than one. He first describes the "process" that Asa Cunningham carried out so well:

*You had to come on freshet water, and you had to be fussy that you caught it just right because you needed all the water you could have to drive the logs. You would put through a modest number of logs and so-called wing-up the river. Once you got it winged, then the logs would go through without catching continuously, because all the places where they would ordinarily catch would be filled with the logs. Then you would go and pick the rear. This was the process. (1402.017-018c)

A particularly troublesome area, early in the drive, was the low banks at Brownfield Meadows. Deering remembers the geography in Brownfield, and the special problems too much water presented:

*You had to be careful, on the Saco particularly, because you didn't want too much water, or the logs would get too far out on the shore and you would have to carry them a tremendous distance back. Each section of the river coming down through Fryeburg and Brownfield, there's a whole area of intervale in Brownfield. One year we had it just covered with logs. We had to carry all those logs back to the river. You couldn't put the logs . . . you'd have booms and stop them at varying places. You couldn't go over a place where all these intervales were on a high pitch of water because as the water dropped away from you, there you were with the logs all dried up. (1402.022-023c)

Deering, when asked how long the drive would last, laughed and said, "Sometimes it lasted two years." He wasn't completely joking:

*We got hung up once on Limington, we couldn't get them off. The selectmen threatened us with a legal process. There were the logs and there wasn't enough water to float them off, so they had to stay there. (1402.017-018c)

Horses would be brought in when lots of timber was left high and dry. On June 16, 1940, Cunningham notes that he "found 80,000 of logs... that we will have to use horses

A view of Turtle Rock in Salmon Falls Gorge in the 1920's now flooded by the Skelton Dam (built in 1945-47). Courtesy Arnold Foster, Limington, Maine.

on." In addition, it was often more than just natural forces of melting snow and rainfall that would wreak havoc with water levels. Deering laid many of his problems at the Cumberland County Power & Light Company's doorstep:

*Logs would dry up [*when water levels were suddenly dropped*] and you might have to carry them fifty or sixty feet. Sometimes two men could lift them, but most of the time it would take eight or ten men with cant dogs hooked onto that log, to carry it from where it had dried up to the water again. Particularly after the power companies came in, they would raise and lower the ponds terrifically. They would use the water as they wanted to, cared nothing about log driving. (1402.018-109c)

Once the drive began, Cunningham coordinated rolling-in at a number of downriver locations. In his 1940 journal Cunningham claims that in the first two weeks, landings were rolled in on the Little Ossipee, Cold River, Great Ossipee and at Kezar Falls. On May 1, he notes that the water came up four feet in the canal (New River at Fryeburg), that the boom was out of Kezar Lake (unleashing all the logs landed there) and that the water was too high to turn over (roll in logs) at Brownfield. Two days later the water was still raging. Bob Littlefield recalls the river running backward, a phenomenon that's hard to believe, but he watched it happen:

*Anyone ever tell you about the condition we'd get down there [Hemlock Covered Bridge] in the spring of the year? What would you think about that river running backwards? To stop it, just wait for the water to go down on Brownfield Meadows. That acted somewhat like a dam. It slowed it up. In the spring of the year the snow was melting, we probably had a good spring freshet up here. The water was running through to Hemlock Bridge [*Old River*] faster than it could run away. It's got to go somewhere. It turned right around and starts running right back in to Lower Kezar Pond, where it should be running out. I've seen it run back in there just as fast as it's ever run out. (1425.020c)

Clarence Brown remembers it as well:

*We took logs out of Fryeburg Harbor, that's further up on—[*the Old River*]. There's a little pond, run right into Old River. In high water, the water coming down the canal [*New River*]. That canal washed through there many years ago. The river used to go over [*the Intervale*] follow what they called Old River. But they had this flood many years ago [*1820's*] and washed that new channel down through. When we was up there in the spring, your logs would go down the canal [*New River*], and there was so much more water going down the canal than there was the Old River, that the water was flowing back up Old River, under Hemlock Bridge and right straight ahead, coming upriver, was the outlet to Kezar Pond. They had to have an awful strong boom to cross the river there in both places, the Old River [*Hobbs Boom*] and the outlet to Kezar Pond. If they didn't, those logs would go back into Kezar Pond. Water was going upriver. They drove both on the Canal and Old River. The logs that was landed in Lovell and Fryeburg Harbor they had to come down the Old River. The Old River and the Canal, a mile below Hemlock Bridge, come together, right into one. (1406.006-007c)

Bob Littlefield concluded:

*We just had to keep our logs up out of there at that time. That's why we had that main boom up where it was. That was up far enough so it was kept away from that. But it's a funny thing, you speak about that with people who live downriver in some of those towns like Buxton and Hollis, they would shake their heads. They thought you were kidding them you know, they wouldn't believe it. One fellow came was up there to see that. He wanted to know when it would be like that. Anytime in the spring of the year, every spring there will come a time when the water will run backwards. (1425.021c)

At the same time that landings were rolled in, pond and lake logs were boomed and pulled to the Saco's entrance, and other crews were at work setting booms downriver. From Cunningham's journal for 1940, we see that the Steep Falls holding boom was set April 25 (remember that Cunningham considered this driving season about one month late). The Hiram boom assembly began April 26 and opened May 22. The Bonny Eagle and Limington booms were set in mid-May. Logs had to be sluiced at Hiram and sorted at Steep Falls. The booms regulated the flow of logs at the Hiram Dam and the Steep Falls Pulp Company, which operated until about 1930. The booms at Limington and Bonny Eagle were fin booms designed to direct logs to one side of the river or the other. Not only did Cunningham need to start rolling in the right logs at the right time, he also had to have virtually all of the booms in place. Steep Falls holding boom had to be set to catch Great Ossipee logs coming into Saco upriver at Cornish. Hiram, of course, had to be ready, but the main boom in Fryeburg had to hold logs above Brownfield. There was also the timber coming down Moose Brook and entering the Saco just above Hiram. The following entries from Cunningham's journal over a month's time show the flurry of activity going on at once:

> Driving Great Ossippi. Rear at the Power Dam tonight. Six men working. Leo Bell and crew at Kezar Lake. Our motor went on the bum and we did not accomplish much. Still rolling landings on the Great Ossippi. Logs are still frozen. We have to cut them out of the ice. Took the motor over to Lewis tonight. (April 29, 1940)

> Rolling in Great Ossippi logs. Rear still at Hell Cat Eddy. No motor and we could not do much with one boat. Put lines on Steep Falls Boom. Saw Mr. Hamilton [*Cumberland County Power and Light*]. He says his booms will be ready to put out in a week. Water too high to put on boards. Kezar logs out of the lake. Water has come up six feet in the canal since we started. (May 2, 1940)

> Finished getting out the Hiram Boom. Water has dropped at Hiram about 18 inches from the high pitch. There is still three feet more to go before we can turn over Brownfield. No men working at Fryeburg today. Holding booms all o.k. Rear at Half Moon Pond. Came down the river the C.C. [*Cumberland County Power and Light*] have not done much at their booms yet. Water seems high at all the dams. (May 7, 1940)

> Driving Old River—Water changed—Bank full and current running out—Running over the banks in some places. I'm afraid it will be bad on Brownfield. Mr. Hamilton told me that he would finish the Bonney Eagle boom today. J.G.D. was up to Old River. C.C. boom not out—Rear of Old River drive at the Old Barn. Put some more lines on Hiram boom—Water is high and logs are piled in—Will be hard to pick out. Three men and a team putting in the Beaver Pond logs at Walker's Bridge. (May 19, 1940)

> Russell Chisholm was drowned today below the mouth of Pleasant Pond. Drivers looked in the afternoon for the body. Sent two men to start logs out of Little Ossippi. Did not do anything on the rear—everybody searching the river. (May 23, 1940)

> Four men on Pleasant Pond boom—Got one over last night—Most of crew looking for the body. Head wind on the Pond today. Motor on the bum. A.C.C. went up at night to tow on the boom. Head wind all night. Boat caught on fire and we had to bail out. Still head wind. (May 25, 1940)

What an enterprise! Cunningham tells us he is driving the Great Ossipee, booming and towing logs on Kezar Lake, preparing the Steep Falls boom, finishing the Hiram boom, and waiting for the right water level to turn over the main

Salmon Falls Gorge with a log jam— prior to 1945. Courtesy Fryeburg Historical Society, Fryeburg, Maine.

At Limington Bridge circa 1910. Left to right: Sam Barry (Bonny Eagle), Herbert Ridlon (West Buxton), Gilbert Usher (Limington), Johnny "Frenchy" Gallant, Ervin Bell (Fryeburg), Everett Boulter (West Buxton), Perley Burnham (Limington), Harry Haley (Limington), Roy Higgins (West Buxton), Ralph Haley (Limington), Lowell Graham (West Buxton), Bill "Father" Hodgeton, Steep Falls. This partially necktied crew was identified by Gilbert Usher on the back of the original photograph. Courtesy of Gilberta (Usher) Ingalls, Alfred, Maine.

boom through Brownfield. Then the water goes up, Russell Chisholm of Cornish drowns, and Cunningham goes to Pleasant Pond to tow a boom at night in the hope there is no head wind blowing the logs back at that time. Wind or no wind, the boat catches fire. This was a complicated, frustrating, and unpredictable business.

Russell Chisholm's demise at age twenty-four is the only work-related death I was told about in the interviews. As Ed Burrill tells the story, it has all the elements of a working man's tragedy.

"It was in the spring of the year, it was cold and the water was high. We had got down to Pleasant Pond, you see the river was high and the river overflowed its banks, and the logs went back into Pleasant Pond. We had a little wing that had caught up on some dead snag out there, and we was going into Pleasant Pond to bring that load of logs out into the main river. We had five or six men there, and one of them was a young guy, he had never been on the river before. We told him, now leave them logs alone there until we come out. Well, I suppose they got cold, and by Jesus they went out and begun to run around and they starting working on them and that thing pulled up and left them. But just before it did them other fellows run back, they got off. But he didn't he was about to ride 'em out. He went downriver there quite a ways, and he could have got off because there was a tree bent over, and them logs went under that tree, he could have got ahold of that tree and got ashore. But he didn't. Well, when he got down a ways, the river began to get a little swifter, and down around the turn there was an old bridge that went out in—years back, some flood, I don't know what year it was. There was a little set of falls there. And he was a good swimmer, he could swim like a fish. And when we got that boom of logs out of the pond there at the river, we went back there and they says, "Russell Chisholm is gone, down on them logs." Well, we took a boat and started right down. And Merle had gone somewhere and had just got back and was coming up the bank of the river. Merle hollered, "Bring that boat in here quick, Russell Chisholm has gone down." And we couldn't believe it. So we went in, and Merle said he went down right over by that tree. See the water comes out around by that point right there and it made right

The batteau in operation at Limington Rips about 1940. This boat was thirty-two feet long and six feet wide at the beam. Deering's batteau was built by the Old Town Canoe Company, Old Town, Maine, in the 1920's. The usual crew had four men rowing with bow and stern men for a total of six. Courtesy J.G. Deering & Son.

over into that shore. I don't know what happened to him, whether there was an undertow there or whatever. I know he was a good swimmer. But he drowned. It was as much as ten or twelve days before they ever found him. Everytime we went to pick a little knot off we were thinking he might be under it. We found his hat, we found that at the outlet of Lovewells Pond. They found him just before you get into Brownfield, that bridge in Brownfield. His brother [*Edmund, just a year older than Russell*] found him. Of course his brother hunted every day for him. But we hunted four or five days, we had to give up to keep the drive going. He got drowned in 1940. (1405.021-022c)

After the landings and the ponds were rolled in, the main body of the drive—some two to seven million board feet of timber annually in the twentieth century (*see table two*)—would begin its downriver trek of seventy miles in four months. The drive had a certain characteristic from Fryeburg down to Hiram—that of keeping the logs in the main river. Ken Blaney recalls the people working the river:

*On the river drive end of it, of course Leo Bell and Roy Smith over in Fryeburg Harbor. Roy Smith was a river driver, and he's very active. And his brother Ken Smith, I think is still living, [*was*] on the river. They had quite a lot of knowledge about that section, and I think they both came down on the drive. Now when they ran the drive down through, there would be some of those men that worked up there wouldn't come clear down through the sixty or seventy miles on the drive. They would go as far as they could go, nicely, to get back and forth. They'd go home every night. The regulars, they called them "white water men,"... started about at Hiram, where Hiram Falls is. From there down through, you had to be a good river driver. (1403.011c)

Ken Blaney remarked that the river became a decidedly different one at Hiram:

*There was sixteen or seventeen regular men that they'd have to hire, that they'd have to have, to take the drive from about where they cleaned Hiram Falls, they called that because the logs would be hanging all over the place there. They would have to go out in the white water and stuff there, and down through, there would be rapids. Just the ordinary man wasn't able to cope with it. He couldn't get out, like myself, I could go along the bank of the river and push logs out, I could do certain things. But they had boatmen, then they'd have drivers that would go right out, they'd have a jam, and they would go right out on those jams and be able to pick the log out and get back ashore. A lot of them couldn't swim, unbelievable, I don't think A.C. Cunningham could swim. They had ways of getting ashore and surviving. (1403.013c)

Frank "King" Graham, Alonzo Foster and Charles Tenney give drive boss Frank Brown a tow in the batteau, sometime before 1924. The photograph is taken from the east bank of the Saco just below Limington Rips at Limington Eddy. The upper boom is hung to hold logs from moving upstream in the eddy. Courtesy of Robert Littlefield, Lovell, Maine.

Seven to ten men would have to work at Hiram sluicing or cleaning logs off the rocks. J.G. Deering and Cumberland County Power and Light Company could never fully agree on appropriate water levels, widths of sluiceways, and ruggedness of booms. Cunningham insisted in 1940 that a larger boom and a wider gap would solve the difficulties. Walter Wyman, President of Central Maine Power, finally agreed that a wider gap in the sluice would be the most practical solution.

Deering kept track of extra expenses from 1933 to 1938, breaking down driving costs in the following categories: above Hiram Dam and over Hiram Falls; foot of Hiram Dam to head of Bonny Eagle Pond over Limington Falls; Bonny Eagle Pond and over Bonny Eagle Dam; Bonny Eagle Pond over Union Dam. The record shows that in April of 1939, Deering's lawyer, Herbert E. Locke of Augusta, totalled damages sought from Cumberland County Power and Light, at $94,500; $16,500 for increased cost of driving and $78,000 for logs damaged after going over the Hiram Dam instead of through the sluice. In August of 1939 a settlement was reached where Cumberland County Power and Light agreed to pay J.G. Deering and Son and the Saco River Driving Company $32,500 in damages.

Moreover, the 1939 agreement acknowledged the right of passage by the driving company and the need to improve sluicing facilities. In addition, the all-important control of water levels was addressed. The victory must have been bittersweet. In 1938, the hurricane effectively glutted the market with timber. In addition, although Deering could not foresee it, the river drive would cease in 1943, just four years later. Deering recalls the whole business this way:

> *I kept fussing about it, and my father says, "Why don't you sue them instead of complaining about this all the time."
>
> The power company had developed, on a retainer basis, most of the good lawyers in the state of Maine. It wasn't easy to get a good lawyer, but I found one in Augusta by the name of Herbert Locke, and I had one in Biddeford by the name of Bob Sidel. And the two of them laid out this suit for the power company and we served it on them. We finally made a settlement with them, and then they spent two or three hundred thousand dollars fixing up the facilities so we could drive logs again. (1402.052-053c)

Charlie Foran remembers what Hiram and other power dams meant to the drivers and to production at the mill:

> *Where all these dams are they had trouble. The last of it they had to hold them and sluice them down through the dam. The power company was using the water, the power company wouldn't let them use the water, when they didn't need to use it themselves. They had these sluiceways, the men would have to be there and put the logs down through there when the power company wanted to use the water. That bothered a lot. (1407.019c)

Stanley (Shorty) Harnett (left) and Merle Cunningham (right) do a little log rolling just below Limington Rips in 1936. Courtesy John Hubbard, Limington, Maine.

Hiram Dam loomed in the minds of the men on the river as well. The response of the men to a very general question—"What were the worst spots on the river, if you had to pick one or two?"—is telling. Leo Bell said:

> *[Hiram Dam]—I guess probably that was about the worst one. Swan's Falls up here, Fryeburg bridge, that's the first real falls. Above, up into Conway there's some quite swift water, no real falls. Now let's see, falls at Hiram. Two sets of falls, Hiram Falls and then below that a little ways there's another set of falls not as high as Hiram, but they called it The Wife, Hiram and his Wife. The next one's down at Steep Falls . . . Salmon Falls, there's one above that, Bonny Eagle, then there's Salmon Falls. (1404.015-016c)

For Leo Bell, the dams and falls were the worst, particularly if the water levels were not set correctly by boards in the dams or if sluiceways were inadequate.

Clarence Brown, who spent much of his time tending the boom at Limington Falls, echoes the difficulty of getting the drive past Hiram Dam:

> *After the logs come over Hiram Dam, then they come down the river themselves, pretty well. When they got down to Limington Falls, I'd come home, stayed home here and worked on the [sluiceway] kept the logs going over Limington Falls. I tended the Falls. (1406.008c)

Ed Burrill tells us another story when asked about the toughest spots on the river:

> *From Bar Mills, there was quite a set of falls there, they called it Salmon Falls. There used to be an iron bridge go across there, and below that bridge was two rocks. The logs would catch on both of them rocks. We'd have to go down, when we got ready, and break that damn jam. The one by the bridge, the first rock, the boat would be there to pick us up. But the next one, the boat couldn't because it was too rough a water. It wasn't very wide in that channel, but it was all ledge up and down it, high ledge. That left that river pretty swift. We'd go out there on that rock, two of us, and pick that jam off. When we got it off, they'd throw a rope out to us. You tie the rope on you, there would be some men in shore, and they would pull you ashore on that rope. That would be the only way they could get you off that rock, not unless they shot you.
>
> Of course, back when we got there, there used to be an awful gang around there, summer people. We'd get left on that rock, and they'd say, "Well

Merle Cunningham navigates a tow boom below Limington Bridge from the Limington to the Standish side. This photo is a good indication of how the river widens out. Courtesy Buzzell Family, Fryeburg, Maine.

how's he going to get off?" [*We said*] "Oh hell, we'll shoot him, we don't need him anymore." They'd believe it. There used to be an awful crowd there. I don't know as you ever seen that before they built that dam [*at Union Falls—Skelton Dam now floods Salmon Falls*]. You wouldn't believe it to see it today. (1405.015c)

*Hiram Falls was kind of bad, but they sluiced them most always at Hiram. The boom broke on us there one spring and let them over the dam, and we had quite a mess. The power company had a regular sluiceway right there, and there was a boom there to stop them from going over the dam. They had a sluiceway at Bonny Eagle, I don't think they had one at West Buxton, I can't remember. They'd sluice them at Steep Falls when that dam was in there. (1405.016c)

Bob Littlefield also gives the first nod to Hiram as the toughest spot on the river, but again, Salmon Falls and a troublesome rock rated a close second. Bob Littlefield also talks about coming of age:

*Hiram Falls was certainly one of the worst. I saw that once when they jammed. They used to run the logs over the dam, when we had water enough. But they plugged right at the lower end down by the power house. We were three weeks getting that jam out of there. You couldn't get out, you couldn't get a crew of men out to work on it. You had to work with a rope and pull them off one at a time. (1425.007c)

*I really thought I had grown up once, when they asked me if I wanted to do that. This was at Salmon Falls, which is underwater now, you know. It was really a pretty wild place. You had to be very careful. There was a rock there, just below the bridge, and the logs would jam against that. They would go through alright, but certain logs, they would keep building up there once in a while. When you let the tail end of the drive through, there would be a dozen or fifteen logs left on this rock. They would be winged in to the shore on the east side so that you could walk out onto them alright, but, if you took the logs off you were stranded out there. What they would do, they would let a fellow go out and take those logs off, toss him out a rope, he would tie it around his waist and hang on, and just jump. And I really thought I had grown up once when they asked me if I wanted to go out and take those logs off. I said sure, I'll take them off. (1425.008c)

Like Burrill and Littlefield, Roy Smith remembered first Hiram, and then Salmon Falls:

Young fellows welcome the drive to the main boom area just above Somesville Bridge. The photographer is on the Biddeford side looking across to Saco. Eighth from the right is Arthur J. Peloquin. Photograph circa 1912. Courtesy of Peloquin family and McArthur Library, Biddeford, Maine.

> *Hiram would be the first place, where you would find them really jammed up. There's a dam where they generate electricity. [*Although there was a sluiceway there*], sometimes, the water would go over the dam there, water would go down, them logs would be hanging, you know. They had a big rope hooked to the batteau, and they would let us right down so the end of the batteau would be sticking out over the dam. I never liked that. I never volunteered to do it, but they always got me into it [*laughs*]. The end would be right out, picking them logs off. It looked a long ways down onto them rocks. (1424.017c)

So the task at Hiram included maintaining the holding boom above the dam to hold the logs until men were there to tend the sluice and nudge logs through the sluiceway over the power dam. Simple enough, unless the worst happened, which it did in 1940, according to Asa Cunningham's journal for May 23:

> Boom broke and went over the dam at noon. It will take two weeks more to get the logs over. I am spending about all of my time at Hiram. It is slow. If we had had a proper sluice way about seventeen feet wide and a good boom I don't think we would have had any trouble.

Part of the difficulty at Hiram for the Deering drive, and for Asa in particular, was the behavior of the Cumberland County Power and Light Company, the owner of the dam at Hiram, not to mention the dams at Bonny Eagle, Buxton, and Bar Mills. Cunningham's shorthand in his journal is the C.C. or the Company. Two days prior to the boom breaking, Cunningham was uneasy when he first saw the boom the Company installed at Hiram to guide the logs to the sluiceway:

> Was to Hiram Dam. The Company boom looks to me to be a very poor one—too small—and I think that we will have a lot of trouble getting logs through that narrow sluice way. Hamilton and Boulter were there. (May 21, 1940)

Everett Boulter and Mr. Hamilton worked at that time for the power company. Ed Burrill remembers Everett Boulter from earlier days.

> *Everett Boulter was a Diamond Match [*woods and drive boss*] man. But along at the last of it he worked for the power company, kind of inspecting along the river. Good man Everett Boulter was, one of the best. (1405.023-024c)

Bob Littlefield remembers Boulter's work for the Diamond Match Company in the early 1920's:

> *Everett Boulter was around then. He used to come up here in the fall and buy stumpage, trees standing, and then they would be cut during the winter and landed someplace where they could be taken out into the Saco River. (1425.006c)

Despite the combined experience of Cunningham and Boulter, it was clear that neither the power company nor Deering would get what they wanted at Hiram. The power company wanted a steady but restricted flow of water to produce power. They certainly would not invest any more than they had to in the boom and sluiceway. They also wanted to keep the sluiceway as small as possible, allowing the least amount of water to pass.

The mention of Everett Boulter reminded many of the river drivers of the earlier days when Deering and Diamond Match ran joint drives. Most of the people interviewed spoke to experiences in later years, in the 1930's and 1940's, after Diamond and Deering stopped the joint drive in the 1920's. Ed Burrill, although he recalls beginning work on the river in 1929, had heard about how the combined drive worked from other drivers:

> *Back in the '20's, they had two drives here, there was two different companies driving. Diamond Match Company had a drive, and Deering had a drive. Well, the logs was all marked, you know, "cross-diamond-cross" was Deering's and "swing H" was Diamond Match. Deering would put on so many men and Diamond Match would put on so many men, and all would drive together. The two crews worked together. When they got down to Saco, they had a sorting gap. They sort that lumber, Deering's would go one side [*east bank*] and Diamond Match the other [*west bank*]. I don't know, they got into some kind of a wrangle there and Diamond Match didn't drive anymore. So Deering had the whole river to himself then. That was back in the twenties. (1405.008c)

Bob Littlefield actually participated in the joint drive because he drove from 1920 to 1925 and then again in 1934. He describes how he liked the river work as an eighteen year-old in 1920:

> *It was fun, for us young fellows. I know that the young fellows, that were mixed in with the older men, you know, it was fun for us, and they would tell how hard it was for them. That it was hard for us to understand, that it was hard work for them.... I started work for [*Frank Brown*]. (1425.004-005c)

Littlefield recalls 1921 or 1922 as the year that he started on the drive working for Deering but ended up working for Diamond Match:

> *Now I'm not sure whether this was the first or second year that I worked for Brown. But, there was one year when Deering had too many men in proportion to the amount of lumber. The crew was divided up that way. The [*company*] that had the largest amount of lumber had more men.... And so, they said we don't actually have too many men working, the thing we must do, Frank Brown came to me and says, "We have got to shift two or three men over, and they will be paid by the Match Company. But they will be working right along in the same crew. If you want to, you can work under me rather than the Match Company's foreman. It's just a matter of book-keeping." So, I shifted over and I worked for the Match Company that year. I guess the next year I went back and I worked for Deering. (1425.006c)

Leo Bell remembers the combined drives this way:

> *I think it would vary from year to year, more or less. Deering might have more one year and perhaps less the next, than the Match Company. I remember one year I drove through for the Match Company and we had sixteen million. Of course they drove together, a Diamond crew and a Deering crew, but they worked together. Probably six or eight men in each crew. If I remember right, if Deering had five million and the Match Company had seven, the Match Company would have to have more help than Deering did. More men, I think. Course that's office work. (1404.011c)

Leo remembers working for both Diamond and Deering, "but worked a lot more for Deering than I ever did the Match Company." (1404.044c) Leo's comment about "office work" illustrates clearly the reality: it didn't make a bit of difference to the men on the river. All that they needed concern themselves with was getting the logs downriver—the sorting of logs came at the end of the drive for Deering and Diamond.

Back on the river, although Salmon Falls looms in Bob Littlefield's mind as "the worst place on the river" (before it was flooded when Skelton Dam was built at Union Falls), there was no boom or sluicing there: "No, you didn't need anything," he said. "It just went and you held your breath that they didn't plug." (1425.015c)

However, there were several spots that needed continual attention, requiring a man or two to "tend the sluiceway." For example, Clarence Brown spent most of his time at Limington keeping the logs moving. He gives us more detail on how a fin boom works in directing logs:

> *They had a fin boom. A fin boom is, you know what a boom of logs is, all hitched together with chains. Well, these were big long logs, and one end, the upper end, is hitched to something on the side of the river they didn't want the logs to go on. The boom logs would go down there and then they had this fin with a brace out here so that when the water hit it, it held that boom, kept it working toward the opposite side of the river. Then logs would come down and hit that boom and they'd follow down.
>
> If they come down on Limington Falls, they come down on the right hand [*west*] side coming down, why they'd pile up on the rocks. They want-

View of Saco River looking downstream 1909. Saco to left, Springs Island Center, Biddeford shore to right. Diamond Match Company's Sawmill is behind right hand distant cribbed pier. The planing mill and box shop are to far right in front of smoke stack. Site was formerly occupied by Saco River Lumber Company. Courtesy McArthur Library, Biddeford, Maine.

ed them all over on the other [*east*] side that they could get over there, that's the reason for the fin boom. (1406.008c)

Although to this point we have moved geographically from Fryeburg to Brownfield and from Steep Falls to Limington and Bonny Eagle, we need to think as Asa Cunningham must have thought: at any given time concentrating on turning over landings on the Great Ossipee or winging up eddies in Brownfield, back and forth, up and down the river. At the same time, he had to think of hanging booms, preparing sluiceways, and setting splashboards. While discussing the boom in Limington, Clarence Brown remembered the fin boom used for sluicing at Canal Bridge in Fryeburg, set with the aid of a capstan raft:

*They had another fin boom down below Canal Bridge to stop the logs from going off into the great big swamp area. We was up there trying to hitch that boom over, and I never worked so hard in my life. Trying to row that boat over, coming upriver against the current. Them logs would hit that boom and go right down under it, there was so much current. (1406.008c)

*[*We used*] a capstan raft, used it in that one spot [*the meadow at Canal Bridge*]. Made up five or ten big logs, drive a toggle chain, on both ends. Drive it into a log and then the other log, roll it up until the chain come tight, put another toggle chain on the top of it so that the logs stayed together. Cut a hole through the middle of the log and put that spindle, which had the capstan, wheel affair.

Hook the cant dog into the hole in the header, which is what the rope wound around, ten inches diameter. Hook the cant dog right into that and you'd start walking right around and round that, every time you come to that rope you'd step over it.

Had to take the anchor out until you come to the end of the rope and drop the anchor and wind it up to the anchor again. Just hooked the cant dog right in. Anything happened, you'd just unhook and let go. (1406.009-010c)

In addition to Canal Bridge, Fryeburg, A.C. Cunningham's journal confirms that fin booms were also used at Limington. On May 20 they "hauled the fins for Limington boom." On June 12, as water levels were going down, Cunningham writes, "We had a big jam in Limington and a lot ran in the dry way." Four men worked steadily into July before clearing the dry way, with the aid of horses.

No book about river driving would be complete without a look at the stories men told about themselves, each other, or their work. Ed Burrill remembered another Cunningham, Merle, Asa's son, who worked the river with great courage. He is featured in many of the drive photographs, a tall husky fellow always in the middle of things.

> *Merle Cunningham. That looks like Merle, right there with the hat on. And he was a ringer on the logs, he didn't fear for nothing... He wore one of those hats, that's just like Merle. (1403.014c)

Bob Littlefield agreed that Merle was fearless to a fault:

> *But he didn't have much fore-thought. He would push into things and it didn't work out. Now, a lot of times there would be these things called wing jams, around Limington Falls. You could take them out a certain way and they would come right out good. Start upriver at the upper end, you understand that these jams would hold a certain amount of water. Well, you start up at the upper end and work your logs into your sluice as you go, and you'll go fine. Merle was one, he would go along to one of those and start down at the lower end, roll two or three logs out, and let the others start and flatten out, and the water would drop and the logs were left in there. He was young, and he just shouldn't have been there. He had the makings of a very good driver, but he would have to change a little bit and listen to some of the old timers. (1425.019c)

Whatever happened to Merle? Surely it would seem that he might follow his father's work on the river. Ken Blaney explains:

> *Merle didn't work [here after 1943], he went in the Merchant Marines [1943]. He was only sixteen when he went in the Navy [1926], he told them he was eighteen. He was in the Navy two or three years, and then he worked for his father. Could never control Merle hardly. Merle was smart and could do a lot. But you know, he was wild. He worked awhile for his father, and worked for Deering, for quite awhile. And it wasn't just his, uh, he liked the water, that's what he liked.
>
> Same as I am, we was about three weeks apart [in age]. He went to be a captain in the Merchant Marine. The last of it, he had a heart problem, and they let him run the ship over across, and he was over in [Norway] when he died [June 24, 1973]. They'd let him go one way and then they'd bring him back, you know, on a plane or something. He had a heart problem, and he knew, before he went, the last trip he made. (1403.015c)

The other thing Ken Blaney remembered about Cunningham was man's best friend:

> *He had a dog Smokey, he was a great fellow for dogs. He used to have a dog when he worked for Deering that used to ride the logs, right on the top of them. I never seen anything like that around, never will again. Big dog that rode right on top of the logs, would go on the drive with him. (1403.015c)

Clarence Brown remembered one of the batteau men:

> *Father Hodgeton he was one of the older drivers. The boys all called him Father. I don't know who started it. He couldn't swim. He used to tend the boat on the falls, take that batteau and come down through in the falls. We had to depend on getting back onto the boat, after we got a bunch of them logs started, they might all flatten out and go. When you was out on the front end there, digging logs and they started in to tumble and fall, you had to get into the boat. He was rugged and he knew how to handle those boats. He and my father handled the boat around the logs, generally. (1406.011-012c)

Ed Burrill remembers, and the photographs illustrate, how logs could jam easily on the many rocks in the river at Limington:

> *One year, I'll tell you a story, we was down here on Limington. Down below the bridge, that is quite a set of falls there, swift water. The logs got hung up on a rock down there on what they called Whaleback. Well, there was Bram Martel, and Merle [Cunningham], and me, we went out there. You know, we took a boat down there, and the boatman, they come up kind of sideways to that jam, which they never should have done, they should have come straight down. It sucked the hind end of that boat right in under that jam and the bow was sticking right up in the air. There we was out there on that thing, and we started working on that, you know, picking it, and it was pretty near ready to go. And they had to run over to Chase's mill and they got a little rowboat and they had some rope and they tied a ball on it and went upstream and throwed it so when it drifted down it hit right where we could get it. They tied that on one end of the boat and they had another rope on the shore end. They pulled us back and forth across. Well when that thing got ready to go, you know, with that little boat, we only had one man out there. We had to leave. They hauled us back and forth there. When that went, Merle was the only one out there. He jumped into that little boat and we pulled him ashore. We lost everything we had, peaveys, raincoats, everything when that went out. Of course that boat went with it, went down into the Bonny Eagle Pond there, and we never got our coats or dinner pails or anything. Whatever we had in that boat was gone. (1405.012-013c)

While the men were on drive, they stayed at a variety of boarding houses, or they may have gone home at night depending on where they were on the river. Clarence Brown remembers the good food, particularly at Frank Hodgdon's house in West Baldwin. Evenings might include horseshoes, card playing, and even accordion playing:

> *They [*boarding houses*] shifted from year to year. The years that I went they were the same ones. They boarded us, we got so much and our board. Got supper at the boarding house, lunch, they brought it to us. When we board up to Fryeburg on the old Holt place, Deering furnished the cook and they'd hired an old house up there and we stayed in that. The other way, when we come to Brownfield, it burned. [*It was the*] Liberty Hotel in Brownfield. Course they were a hotel, and they furnished the help there.
>
> Frank Hodgdon's was in West Baldwin. We stayed there, it was one of the best places we boarded. It was just like home, and boy did they put on a feed up there. Beautiful.
>
> We'd go so far on the river, and when they got so far—they called it the rear, that's the tail end of the logs—when they got so far, why then they'd move to another boarding house. (1406.012-013c)

We asked Ed Burrill about leisure time activities and he remembered log-rolling contests just after they moved the drive over Salmon Falls. When asked if there was any singing at the boarding houses at night, Ed replied, "No. Only when we got a little of the Oh-Be-Joyful—we got a little of that once in awhile." (1405.043c)

Bob Littlefield recently recalled another event at Salmon Falls reported in the April 1992 issue of the *Fryeburg Historical Newsletter*.

> One unusual event on the down river run occurred at the Salmon Falls bridge. Kate Douglas Wiggin, a Maine author of children's stories who wrote *Peabody Pew* and *Mrs. Wiggs and the Cabbage Patch*, had married an Englishman by the name of Riggs. Her original home was in Salmon Falls, but she had spent her later years in England. Bob was driving logs at the Falls when men stepped out of two cars on the bridge and asked the crew to pause in their run while they sprinkled the author's ashes over the bridge to comply with her last wishes.[12]

And now time for a story from Ed Burrill:

> *Well, I want to tell you about our bank we had on the river. When we got down on Limington, there is a big rock down there, it had a big seam in it. Each one of us river drivers would go and put a penny in that bank. We would take our peavey and drive it in by that seam and it would drop down. Every year we put a penny in there so we wouldn't be broke.
>
> Well, [*Deering*] had a bunch of guys come from Bangor. After they got done on the drive they went out there with a rock chisel and chiselled that rock out and stole them pennies. So they robbed us. I don't know just who it was that done it, but I laid it to them fellows. They robbed us. (1405.043c)

Ken Blaney has a more supernatural story:

> *I can remember on the drive right down here below Cornish [*puffs on his pipe*]. The drivers were down on the river, and A.C. and I came over. There was a big pine tree there, and we sat down for a minute under the tree. Then come this big lightning storm. A.C. said, "I've got to go out and see those men out there. You want to take me out there?"
>
> We just got out there and I looked back and that, the only time I ever see lightning strike a tree, struck that tree we was sitting under, took the top right off, and come right down the tree and went out in the river, and those fellows had caulked shoes on, and did you ever see a sight, those fellows jumping. They jumped that high, and the shock, whatever it was, you know, electricity went into those caulks and right up through. Of course it was wet, and it was a comical sight to see those fellows jumping. Not comical really, but—the only thing that saved our lives was the fact that we just got up and went out there, because if we'd been under there when that come down. Just one of those fates. (1403.019-020c)

The drive was complete once the logs got to the holding boom above Saco and Biddeford. Charlie Foran describes the boom at the Biddeford Mill, and the upriver holding and sorting booms:

> *We used to keep about a week's supply [*of logs*] down there [*at the mill*]. We didn't dare to keep too much, because if that let go they went out in the Atlantic Ocean, right down over the dam. Up above, we kept more, up top the railroad bridge, about a mile up... another bunch about a week's supply. And then, way back, we kept the main boom, a big supply. The men would go up there, they would let a bunch down, they would come down by the railroad bridge. We didn't have them down here because if the boom let go they would go out in the ocean. That's why we kept them up there. (1407.056-057c)

It was a drive that required perfect timing in rolling the logs in on the many tributaries and lakes upriver that had Deering timber. At the same time, farther south, Great Ossipee and Little Ossipee logs were rolled in to give the Deering sawmill material to work with as early as possible in the year. Hiram Dam and Salmon Falls were clearly the most challenging physical obstructions to the drive. Yet all along, it was a case of keeping the timber moving, picking off jams, and sluicing logs through the clearest passage.

J.G. Deering & Son sawmill, built in 1914, as viewed from Saco. The photograph was taken in the late 1930's.
Courtesy of J.G. Deering & Son.

CHAPTER FOUR
The Sawmill

Background

Having traveled the course of the river drive from the Maine-New Hampshire border over a hundred miles to Saco-Biddeford, the logs have arrived at Deering's lumber manufacturing operation. Although a good half dozen people had something to say about the sawmill, portable mill operations, trucking, and the retail store, one person tells us, in great detail, the technology involved in manufacturing lumber.

Charlie Foran was 87 years old when we interviewed him in September 1980. In the truest sense of the phrase, he was Joseph Deering's right-hand man for operations. Charlie Foran began working for Deering in 1920, about a year after Joe Deering came back from World War I to start work in the business with his father. Charlie drove a truck for four years and then spent nine years in the retail department of the Biddeford store. From there he learned how to run the saw and planing mills. He became yard and mill foreman in 1933 and continued in that capacity until the sawmill shut down in 1948. Charlie recalled how he happened to move from the retail department to the sawmill:

*Then, one winter, well, [in the] winter time I used to help overhaul the trucks and cars. After one winter, Deering wanted to change steam engines to run the mill. He bought an engine from the Diamond Match Company, and they were going to have to overhaul it. A man from the Portland Machine Company came to overhaul it, and they had a fellow to help him. But one day something happened, and they put me to help him. His name was Charlie, too, Charlie Thom. He took a shine to me, you know. The next day they gave me another job [and got] another fellow to help him. (1407.005c)

At that time, Joseph Deering's father, Frank, would check in at the mill each day to see how things were going. The mechanic from Portland Machine Company asked him where Charlie Foran was and made clear to Frank that he really wanted him back as his helper. In speaking to the yard foreman, Frank said the Portland mechanic "wants Charlie back. He only worked one day, he wants him back, so you go get him. You leave him here, don't you take him off." (1407.005c) So Charlie Foran went back to work with Mr. Thom for most of that winter:

*We had to re-bore the cylinder in the engine with a hand operated machine. We did the valve and the valve cylinder. They did the machine work on the piston, the valve in Portland. We put it together and I helped him fix the bearing. In the meantime, we had to start up the engine and I learned to run it. It was a steam engine. I used to run it an hour or so at a time. [*Then*] we'd shut it down and see if it would heat up anywhere. (1407.005-006c)

Charlie learned how to rebuild the engine and how to run the mill. Charlie Thom from Portland Machine went to the Deerings to say that Charlie Foran should be the one to break the engine in when the sawmill began operation. Foran stayed for a week or so watching and lubricating the new engine and then went back to the retail department. He did other things that year, 1932, including driving Joseph Deering up the Saco Valley to woodlots, and, when a teamster took sick, he worked with the horses driving scoot. The following spring, the yard and mill foreman, Mr. Roberts, retired. Joe Deering told Charlie he had the job if he wanted it. Charlie said:

*I was shocked, to tell you the truth, you know. During the summer I had a job, I was the yard boss. We had to manufacture the finish too [*in the planing mill*]. Mr. Roberts took care of the sawmill and I took care of the finishing, see, to make it easier on me. So, that fall he retired and I had it by myself. Well, during the winter, every winter, Mr. Deering used to overhaul the machinery, so when it came springtime and summertime, he started the mill and it didn't break down. He wanted it in A-1 shape. They did that, and in the spring I took over, ever after. I had charge of the manufacturing, the finishing, and the shipping. (1407.007c)

What follows is Charlie Foran's amazingly precise description of how things worked from the moment a sawlog came out of the river until it left town by rail or truck or was sold in the retail store. Admittedly, we are relying heavily on Charlie's memory of work he did thirty years earlier. Others, such as Ralph Morin and Phyllis Deschambeault help confirm details of how the Biddeford mill was set up or how it worked. Chet Leonard, although he only worked on portable mills, could tell us how sawmills worked whether stationary or portable.

The Sawmill

Charlie Foran told us how the sawmill worked on two separate occasions, on September 16, 1980 and again on November 19, 1980. Based on the first taped interview, my wife and I drew a sketch of the floor plan of the Deering sawmill (on the second floor) and the planing mill (on the first floor). Charlie's narrative in this book is a combination of both sessions. In some cases, his description proved clearer in the first interview, while at other times, perhaps because his memory was prompted by our sketch, he went into more detail during the second interview. He first described how the logs were moved from the river into the mill:

*There's the mill, that [*log chain*] went right up from [*the slip on the river*]—right up in the air... [*at a 45° angle*] into the mill. The slip was about, oh, three feet wide. The chain was in the center, and we had a walk to go up onto the side of it... When that log got up into the mill, that chain went along the floor of the mill. The logs went along and we had a brow there, we called it, to store the logs before they went on the carriage of the mill. At first we used to have a man roll them off with a cant dog and if he didn't have time to roll it off, we had a lever there and that log hit a bumper and the chain stopped so he could do it. But we tried to train them quick enough so that we could keep it going. After awhile we put in a jack. All you had to do was step on it and the arms came up through, the logs rolled off themselves. They rolled down this little platform, we had to just keep them straight. We had some arms down at the other end that stopped them like this. It would hold them until he [*the sawyer*] wanted a log on the carriage. (1407.044-045c)

Ralph Morin worked for Charlie in the sawmill beginning in 1934. "I used to do this quite a few times—feeding a slip... This is where the logs come in, in the river with a pick-pole you'd put it onto a chain. A big, long chain going up, then they would bring the logs right up..." (2010.012c) Charlie Foran continues:

*A man there at the river with a pick pole, he pushed the logs on there. When the fellow had enough upstairs, he pushed the lever to stop it until they had space for more. They would saw three or four, or four or five, and he'd stop them coming up until he needed more. (1407.045c)

Charlie Foran describes the process once the log is in the building. The carriage track of the sawmill was sixty feet long, more or less, and it took up a good half of the building. *"We had a big bumper on each end with an air compressor so if the carriage got away it hit that bumper." (1407.050c)

*They stored those logs, they had space [*a platform*], [*at*] the end of the building, oh twenty feet... Only the track, the track of the sawmill went up by... So when they wanted another log, that carriage went up by this platform, and when it got even with the platform they loaded it [*the log*] on. (1407.146c)

Once the log was on the carriage, it was clamped into place and sawn lengthwise. Once a slab was taken off one side, the "dogger" and the "setter" would turn the log to take another side off, eventually ending up with a squared length of timber. Charlie Foran explains:

*They had two men working on the carriage. One was what they called the setter, he's the one

White Pine on the Saco River

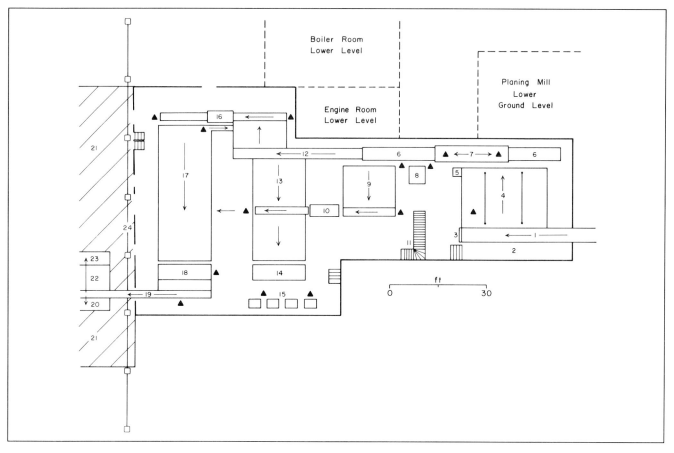

Sawmill Diagram

J.G. Deering & Son Steam Sawmill Biddeford, ME built in 1913, replacing a steam sawmill (probably built 1893) that burned on February 6, 1913. Framed with 12' X 12' timbers - this mill was the ultimate for its day. The mill floor at second level of the building was 135' x 36'. Subsequent additions widened the Elm Street end to 68'. The floor is of 2" spruce/hemlock plank over 1" subfloor. Ceiling heights vary from 12' to 16'. The filing room (for saw and knife sharpening) at a third level was located directly above the bandmill for ease in changing saws. The floor plan as shown is appx. as it functioned in the 1930's until 1948. Additions to the 1913 plan accommodated a re-saw, new trimmer, slasher saws and an overhead crane on the green lumber brow. Lacking photos or precise records of the machinery, the layout and specifications are drawn as recalled by Charlie Foran, Ralph Morin and other men who worked there. Power was generated by a steam engine fueled by three boilers burning wood waste. All machinery (except for several small electric motors) was driven by leather and fiber belting running off steel pulleys and shafting. Charlie Foran recalls the record sawing of 53,648 bf for one shift in 1938. The 7.5 million bf from the 1940 drive was sawn in five months by running two shifts 5:00 AM to 9:00 PM - 4375 bf per hour average. Late in the fall of 1948 the mill saws fell silent for a final time. J.G. Deering & Son was, to our knowledge, the last sawmill located directly on the Saco River - from the White Mountains to the sea - to manufacture lumber. Literally hundreds of mills, large and small, had come and gone along the Saco since the first recorded mill privilege - Richard Spencer's of 1653 in Biddeford. On that final day in 1948 - signaling the end of 295 years of continuous lumber manufacture on the Saco - Charlie Foran invited Irene C. Maher - then a company clerk, and today Office Manager with 50 years of service to the Deering Company - to pull the 4:00 PM steam whistle, signalling the end of an era. The 1913 building still stands today, gray-bearded yet solidly timbered. Obsolete for its intended original use, the mill continues in service for limited lumber and building materials storage.

Key:
1. conveyor log chain from Saco River log slip below - approximately 75' away
2. walkway
3. log bumper
4. log deck platform or brow sloping to the carriage - graduated above the floor from approximately 30" down to 18"
5. steam piston "log kicker" for pushing logs on to carriage
6. 60' track and steam works for two directional carriage with shotgun feed
7. Clark log carriage - approximately 4' X 25'. Two men rode carriage to operate setworks and "dog" the logs
8. 7' Clark R.H. 10" bandmill
9. conveyor chain at waist height carrying lumber to be edged
10. three saw Lane edger
11. stairs to third level filing room 29' X 36' - with overhead access to bandsaw
12. live rolls to re-saw
13. conveyor chain for slabs and edgings at floor level under edger
14. five saw pedal slasher for trimming slabs and edgings to 4'
15. waste disposal - edgings in two hoppers to left, slabs in third, other scraps to fourth hopper to be ground up in a "hog" below. About 1940 a large chipper was built near mill for more of this waste
16. single 8" band re-saw installed about 1933 for splitting plank and cants
17. conveyor chain approximately 18" off floor carrying lumber ready to trim and grade
18. eleven saw pedal trimmer for trimming 6' to 16' lengths
19. "surveyors" table to mark and tally each piece
20. grading table - lower grades moved to left (river side) piles
21. large uncovered platform (cross-hatched) 55' X 80' - approximately 4' below sawmill floor - known as green lumber brow. Plan shows only the part of this platform nearest mill
22. short conveyor chain for higher grades carried lumber through anti-stain dipping tank
23. sorting table - higher grades
24. overhead crane 32' wide X 120' long parallel to mill wall. Crane hoist positioned finished stacks of lumber for take away by wagon or truck. Plan shows only the concrete piers nearest the mill that carried this structure. (Built about 1934-35).

that gauged the thickness that they was going to cut off. And the other fellow was what they called the dogger, and he had two levers and they had like pig's feet to fasten the log on with. They had another lever that they had two jacks on the carriage where the log went up against. This other lever, he squeezed them and little prongs caught the logs like this to help hold it. [*Then*] they started to saw.

The man that operated the saw [*is the sawyer*]. We had a track, you know, the carriage went on a track. He would take the slab off, then he might take one or two boards off, and he would turn the log over until he got four sides all clean. He used to cut the log down to eight inches, that's as far as he went. What I mean, he kept that log until he got eight inches square. After the eight inches square he would split it [*and drop it down on live rolls*], and we would send it over to the re-saw and they would do the rest. See what I mean? He'd cut it down to eight inches, send it on to another guy. If it was a little small log he would just slab it off and let her go. If it was a big log he would cut it down to eight inches. And then he would send it over to the re-saw. Everything went on live rolls, moved right along. (1407.046-047c)

Ralph Morin confirms this method of squaring up large logs with the headsaw, then letting the re-saw finish the job:

*That log rolls right down towards the carriage. Then the sawyer is the one who handles the carriage... They would cut the log right there. They'd make slabs and stuff like that. Then he'd probably leave a piece about six or eight inches thick square and [*after splitting it*] he'd let that go straight down the other side of the building... to what they call a re-saw. That re-saw would have three men on it, one operating the saw and the other two men help bring in the lumber, one on each end. (2010.015c)

The Deering mill bandsaws were ten inches and eight inches wide. Charlie Foran remembers it this way:

*Bandsaws, yeah, we had a ten inch bandsaw to cut the logs ten inches wide. Forty one foot six [*inch*] endless length. And we had an eight inch bandsaw [*re-saw*] to split it up into one inch rafters. And that was thirty-one foot six, endless length. And his bandmill was what they called a seven foot mill. You had two wheels, seven feet in diameter, top and bottom. You cranked and loosened them up and put the saw in, and then you cranked it up, spread the wheels, tighten it up. Everything went by steam, he used refuse fuel, shavings, or ground up stuff. We had a hog to grind up the stuff. When we had a lot of shavings, we didn't grind so much. (1407.015-016c)

The re-saw occupied an alcove-type space at the end of the building. Depending on whether the board needed to be edged or trimmed, either chains or live rollers were there to take the boards to the edger or the trimmer, which was on the Saco side of the mill building. Charlie confirms Ralph Morin's recollection that it took three men to re-saw. He also goes on to talk about edging and trimming:

*We called them the sawyer and the re-saw fellow. The re-saw fellow had two helpers, the one on the back helped him to tip up the plank on the edge to put them through. The fellow on the end took them away, he tipped them over and let them go on some chains. (1407.016-017c)

*If it had bark on it, we had an edger machine, up in here on this side of the saw. That fellow would step on a pedal and the stop would come up and it would stop there and it would come back up here. We had an edgerman there and a helper, if the board, [*after squaring was*] eight inches, why he would set his saw to cut eight inches or seven or ten or whatever it was that would square off. Then the stickings used to fly off on the floor. But later we put a chipper in there and we let the stickings go down there and down in a chute and we ground them up, so we didn't have to handle them. The slabs went down on these chains underneath these sets of rolls and went out into another shed that stood right out in here on the side, connected to the main building. They were overhead and we had chutes underneath, tapered down. We'd open the trap and let them fall into a truck or on a dumpcart, years ago it was dumpcart, then we had trucks in there after awhile. (1407.048-050c).

*Then we put in a rig to grind up the stickings, to make chips. We used to grind them up and load a car of chips a day. He sold them. During the war he sold them, they made roofing paper, they made dynamite flour for the army. (1407.016-017c)

*We had a man that used to do the trimming, he had several saws there. He had pedals like pedals on the piano. These saws were all on weights, and he had a stationary one, and if he wanted to cut the board ten foot long he stepped on the ten foot saw and let it come up, cut that off. He'd take his foot off and the saw would disappear. He could cut his edges off mighty quick. If he cut off too much, that's waste. That's where I had two fellows there for awhile. Circular saws. The only bandsaw that we had was the big saw that cut the logs and the saw that split up into boards. Those were the only bandsaws that we had. (1407.016-017c)

To recap, then, logs came up from the river into the mill via a conveyor chain. The brow or log storage deck sloped towards the carriage. Each log was then positioned one by one on the carriage with the assistance of a steam piston "kicker." The carriage was next activated—past the bandsaw—for the number of cuts judged necessary by the sawyer to reduce the log to the appropriate size(s). These pieces, if in final form, went directly to the edger or—if still in timber or cant size—to the resaw for further manufacturing. All lumber passed finally by the trimmer and on to the surveyor and grading platform. After trimming, Charlie Foran recalls:

> *The boards went on to a set of rolls and the surveyor stood out here in the alleyway. We had a door that went outdoors there, and the surveyor stood there and he marked the boards, how many feet there was in it, and he kept track of how many feet there was. And the boards went outside and we graded them outside on a big platform. (1407.053c)

Ralph Morin's father, Hilaire Morin (they called him Eli) worked for J.G. Deering and Son for more than fifty years. He surveyed lumber as it came out of the mill and stacked it, and later on he operated the trimmer saw. Ralph remembers both of his father's jobs this way:

> *The boards would come towards the wall... This is where my father would take the board, look at how long it is, with his feet he would lift up the saw that he needs for the length of the board. Then he'd let the board go and the chains would take the board through and they'd cut the board to length. Then the board would fall onto another table right up against the wall here. When they fall right there, they had a man marking, putting a mark on it—how many board feet is in that board... then he'd pile it up out here, three or four boards and he'd just give it a shove and they would roll down onto the brow. We call it a brow. The brow is outdoors. Once it got outdoors, then you had one man grading it. He probably had six or seven men to grab the boards and put them to their grading, where they belong according to width and grade. Then when the load got big enough, this is when the man with the [*electric overhead*] crane would take over. He would

View of J.G. Deering & Son brow and carriage in the sawmill circa 1925. Left to right: Fernand Roy, Biddeford, "Roller" — Mr. Cleveland, "Sawyer" - Unidentified, "Millwright" — Unidentified — Unidentified, "Saw Filer"—Unidentified, Ernest E. Whitten, Saco, rode carriage as "Setter." Courtesy of J.G. Deering & Son.

load it on the wagon and then they'd take it up to either Scammon Street yard or wherever it went in Saco. They'd let it go there for about a year to dry, air dry. If they needed certain boards, that's where they would go and that's where it would come back to the planer mill to be planed, to be dressed. [*The lumber would be planed*] after it is dry, yes. You cannot plane it when it is green. (2010.016-017c)

Before moving on from the sawmill to the planing mill, Charlie Foran describes one other piece of work that took place in the third floor of the mill building, an area known as the filing room:

> *And we had a flight of stairs that went up in the attic to the filing room where we sharpened the saws and things…. We had a trap door, when they changed saws…, you stopped the mill and [*untighten the*] wheels. You would let them loose, we would take the saw off and we would heist it upstairs and they would put another sharp one down through and they would put it on and tighten the wheels up again. Oh, we had, every spring we bought four brand new ones. I would say that we had a dozen ten inch bandsaws there. [*MC: Did you use that whole dozen?*] No, some of them would get cracked, they would get worn down, or we would have a break, seven or eight teeth off them. If they only had one or two teeth, why, he would let that go. We used it on hemlock or something where it didn't saw too smooth, we didn't care, because you planed it off. But on pine board we didn't use it, good boards. Then as the saws got narrow, we used them on hemlock, because we kept the good saws for the pine.
>
> [*MC: How often were the saws sharpened?*] We changed, if we didn't strike any nails or anything, had good luck, sometimes we ran the saw half a day. But usually we used to change at 9:00 or 9:30. We started at 6:30 in the morning. We changed 9:00 or 9:30, then we changed again during the noon hour, and sometimes… at 3:00 again.
>
> We had a man upstairs full time, that's all he did, work on the saws. He put temper in them and he sharpened them and did everything. And the re-saws too. The bandsaws were 14 gauge and the re-saws were 16 gauge, they were lighter weight. Then we had these saws that went on the edger machine, then we had saws that cut up the slabs, the stickings, see. This fellow used to have machines that sharpened all of those. Those were circular saws. Then we had a man, he used to temper them, too. He welded them and everything. [*If*] it had a crack in it he would weld it. (1407.053-055c)

Ralph Morin remembered the pace of work in the filing room:

> *See this saw right here at the top. They could drop that saw right down in there and then it just rotates around this machine. From this machine is where they have this automatic filer.
>
> [*How often they have to change a saw blade?*] Well, I would say they'd have to change it every day. Of course once in a while they'd hit spikes and stuff like that. A nail wouldn't bother it, but a bigger spike or even rocks inside of the claw then they'd have to stop immediately. (2010.036-037c)

The sawmill would operate as soon as weather and log supply (local or driven) permitted continuing until ice covered the river in the winter. Charlie Foran was ever so conscious of finishing the sawmill work:

> *The sawmill, the river would freeze up, liable to freeze the last of November. We had to saw the logs by the first of December. So we used to run the sawmill ten hours a day, only worked one shift. So we'd be sure to get the logs done. But when we ran two shifts, they [*each*] ran seven hours a day, why we didn't have time to put the booms away, they froze right up the next day. (1407.018c)

He was also conscious of the production level of his mill. For example, he wanted to know how many logs it took per thousand board feet of lumber:

> *They used to count the logs, the man at the river, he would put on a piece of board every log he put up, he counted them. And the man at the top, he had to count them too, to check one another, to see. That's how we determined how many logs it took for a thousand feet of lumber… That's how you could get your average of how many logs it took [*a*] day… Just as the fellow down to the bottom had a piece of board, and he'd mark one, two, three, four, five. The fellow up at the top did it the same way. They turned those in everyday. (1407.059-060c)

He also wanted to know, for his own information, how many logs the mill could saw in a minute:

> *We had a door up here where the logs went in… They had this bed that was about twelve foot wide that we filled up with logs to have them ready for the sawmill. We could saw, for ten hours, a log and a quarter a minute, on the average. I kept track of that for a long time. I didn't always tell Deering that, but I kept track of that. We averaged a log and a quarter a minute, for ten hours… There were days that we sawed a thousand logs in a day. There were days that we sawed 600, according to the size of the logs. As they came down the river, we'd have a good bunch of logs that were big, then we'd have a bunch of small ones. (1407.056c)

Filing room of J.G. Deering & Son sawmill, circa 1925. Courtesy of Ernest E. Whitten, Saco, ME, first on the right.

The Planing Mill

From the sawmill, after appropriate drying time, the lumber went to a separate operation on the first floor: the planing mill. As in the sawmill discussion, I asked Charlie Foran to respond to a floor plan we sketched of the planing mill based on the first interview (*see page 62*). Charlie was firm in his view that the sawmill and the planing mill were two separate operations:

*We had three planers. We had one planing machine that planed either one side or two sides. And we had one planing machine that would make matched boards [*tongue and groove*] or plane four sides... Then we had a machine that would make mouldings. We used to make all kinds of mouldings, change the knives and make different kinds of house finish. Then we had a little bandsaw down in the planing room, we used to take boards and make clapboards out of them. Then we had another workman in the planing room, if somebody wanted in the retail department a board just exactly seven inches wide, [*he would*] take an eight inch board and cut to seven. We had a sanding machine to sand the lumber, the finish lumber, one side there... The sawmill and the planing mill were separate from one another. (1407.017-018c)

Ralph Morin, hired by Charlie Foran in 1934, began by working in the planing mill. He remembers the difference in the pace of work between the sawmill and the planing mill:

*They had a big [*steam*] engine that was strong enough to run the whole outfit: sawmill and [*the planing mill*]. When they closed the sawmill down, they switched to a smaller machine with a smaller engine and they would run the planer. The planer mill was a twelve-month a year deal. (2010.010c)

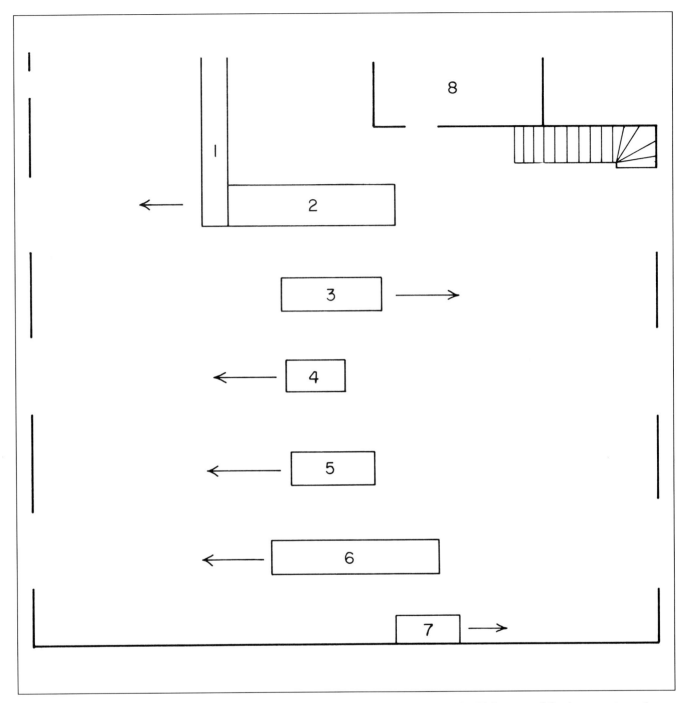

J.G. Deering & Son Planing Mill Biddeford, ME. Built in 1913, this is the floor plan (at ground floor level below sawmill floor) appx. as it was from 1935-1958. Dimensions 48' X 60', usual work force 4 to 6 men. Scale 1" = 8'.

KEY:
1. Trimmer table with manual swing saw (replaced by a more automatic trimmer outside building in the lumber yard early 1940's).
2. J.A. Fay and Egan — Four side planer — Six Knife Babbett Bearing.
3. Berlin 5" bandsaw for resawing clapboards and planks.
4. Berlin surface planer — two knife— Babbett Bearing square head — 12" X 30" maximum opening.
5. Yates-American bench rip saw (removed to cabinet shop about 1950).
6. S.A. Woods 12" Molder/Planer — Babbett Bearing.
7. Yates-American 24" sander (removed to cabinet shop about 1950).
8. Sharpening and repair room.
   Arrows indicate lumber flow directions. Layout and specifications per Charlie Foran, Ralph Morin and others.

The mention of using a smaller engine to run the planing mill raises the question of how exactly the mill was powered. Ralph Morin, looking at photograph on p. 65 explains:

> *This is what they call trimming. Remember I said my father cut the ends of boards trimmings? Well these are the trimmings. It would feed the boiler. The shavings from the planer mill would also run into the boiler room. This is a door entering into the boiler room right here. That's Ed Corney... We called him Ed so it must have been Edward.... He was right next to Mr. Deering.... He was probably like a vice-president. He was right next, then Charlie [Foran] was next. These little doors here, on the right of the photograph, we used to pass lumber up there over the boiler, over the dry-kiln. That was a hot job up there.... Nobody wanted to go up there—not even me, unless we had to. It was hot stuff. [*The skylight is*] right over the boilers themselves. Your fires burned right beneath this part here. These here were all shavings going directly into the three holes where the fires were. [*MC: What are the four funnel shaped affairs?*] Shavings were blown in there from the planer mill. We needed shavings. Shavings are dry. The boards when they're planed are dry. I helped them in there a little bit. I helped Oliver Lambert. He was a very good fireman. When we needed a little better fire, well, we'd open up some of the shavings and let the shavings drop in order to help burn these green pieces of wood. All the saw dust from the mill went down there. It was to heat up the fire, but it was all shavings in there. Mr. Lambert would come in there around four o'clock every morning to start his fires up. (2010.020-022c)

Not only did Morin describe the fuel for the boilers, he also described a makeshift method of quick-drying green lumber near the boiler.

According to Charlie's description, the planing mill machinery sat in the middle of the first floor of the mill building—seven machines in a row crossing the building. The Yates-American sander was next to the wall on the Saco River side of the building. Next to it was a twelve-inch moulder and planer made by S.A. Woods. The third machine in line was a Yates-American rip-saw. Next to that was the big thirty-inch Berlin surface planer:

> *We had a surface planer, planed one or two sides, whichever you wanted. That was a good sized one, planed twenty-four inches wide, thirty inches wide, that was a Berlin machine... It was a square-headed machine, old- timer. You could plane a twelve by twelve [*timber*] on it, could go down twelve inches. (1407.061c)

Next to the Berlin planer was a five-inch Berlin band-saw for manufacturing clapboards. Next to that:

> *We had a J.A. Fay & Egan planer, big one. It was a four sider, we could put the bead [*pattern*] on the boards or anything. That was a four-sided planer, this molder was a four sided, because we could plane molding or make molding or anything on [*it*]... We had a trimmer saw, table saw to trim, cut the end of the boards off. See, the board went this way, and we could, swing saw, we could trim the ends of the boards off by hand.
> [*Were all seven machines in a row?*] Yes, all side by side.... practically in the middle [*of the building*] because, see, we had doors on each end, we planed anything, molding here, why we could drop it on the floor or we could load it on a wagon. And then the middle planer, we could take it out the door. (1407.062-063c)

The rip-saw was next to the S.A. Woods molder and planer for ease of preparing lumber being made into molding:

> *And when we wanted to make moldings, we'd bring in a load of boards and dump them in the back here. If we wanted to rip them up, we were going to make a molding two inches or three inches wide, we would rip the boards up three inches, [*if*] we didn't have any boards three inches wide, then we'd rip them up and put them through the molder and make molding with it. (1407.063-064c)

Lumber was graded as it came out of the sawmill, before it was hauled to the Scammon Street and Boom Road yards for sticking and drying, and it was brought back at a later date for planing. Grading took place on a platform at the south end of the building. Charlie Foran remembers the process this way:

> *We had one man that did the grading, see. Then we had five men that took care of [*grades*]. We had three on one side and two on the other. The table, one side of him, he'd stand there, Vernon Waterhouse was the last one. He'd stand there and he would grade it and he would put the #4 and the rotten on that side [*the right*] and the #1 to #3 on the other, #3 or better on this side. We used to dip the #3 or better in lignisan [*preservative*] in a tank. We had a tank there, a V tank, it had chains there going into it and out. And he would just tip it over and let it go into that tank, it would come out on chains, it would drain a little bit and the men would pile it up. So it wouldn't turn blue, blue stains, see. If it did turn blue it didn't sink [*deeply*] with that lignisan.... But if you didn't [*treat the wood, stains*] would go right in through the board, that blue. Lignisan, yeah, we used to buy it [*as powder*] from the DuPont Company. (1407.067-068c)

From there, the boards were "stuck up" to dry in the board yards. Green, unseasoned lumber required the lignisan treatment from May to September to avoid excessive "down grade loss" from the blue stain that would appear in an otherwise #1 grade board. Down grading lumber meant losing money.

Once the lumber had dried, for six to twelve months, it was brought on an "as needed" basis to be planed. Ralph Morin's first job was taking away lumber from the planer. After about a year, he began work outside as a surveyor:

> *My first job was taking away from the planer. From there I asked Charlie to get out of the planer mill and he said what do you want to do? Well, I said I would like to go outside and work. Of course, I had no high school education, but he said, I know you can count. So he sent me out there loading boards and he taught me how to read the figures on them and add them up so I became a surveyor out there.... It was just in the summer time, because in the winter time, they kept their [senior] crew members. It was just planing in the winter time. The saw mill wasn't running so they kept the oldest people in there. In the winter time again, I go back in the woods chopping wood. (2010.004c)

The planing mill kept certain kinds of lumber in stock at all times—whether it was clapboards, matched tongue and groove boards, or beaded molding. Ralph's work evolved in the 1930's from the planing mill to being a lumber surveyor in the yard—tallying production or preparing orders for shipment. But Ralph's job would change again, by his own choice, because of demands in the lumber business brought on by World War II:

> *We would keep some in stock all the time. They keep a number four grade; they would keep probably eighteen, twenty thousand feet in a pile at all times. This is for people coming off the street and of course retailing. If they order a matched [board], you need to have some ready. In the winter time, at the beginning the first winter I started there, most of my work was loading for shipping, for car loads. Railroad cars.... There was a lot of wholesale [back in the 1930's]. Then life seemed to change after the war. Then they started going into retail much heavier. This is when I decided to make the move [to retail]. I said to the man in charge there, I'd like to get into retail if I can. By then I was surveying outdoors—counting lumber. I wanted to get into retail because I liked being with the people. Finally they gave me a job in retail, that is driving truck. During the war, I stayed here all through the war years because I was driving a truck. They gave me deferment... A lot of mornings I'd leave at four or five o'clock in the morning. I'd go to the Navy Yard in Portsmouth, dump the load, come back and then we'd have to load. I'd start in Biddeford Pool and I'd go around Biddeford Pool, Kennebunk, Kennebunkport those places, as far as York Beach and Sanford and this was door-to-door delivery. Then the next morning there was another load ready to go either to the shipyard [South Portland] or to the Navy Yard. (2010.018-019c)

> *I drove trucks through the war. Then in 1949, Martin Lee—I think he had just replaced Charlie [Foran]. Anyway he was the head of it then. He asked me to go out on the road to sell. I said, Oh geez, not me. I'm not a salesman. He said I want you to go out. I said why. Well, he says, You're driving truck, you're coming back with a few orders, which I didn't have to but people say can you send me this tomorrow? So they sent me on the road. I was on the road for eighteen years. (2010.028- 029c)

Wholesaling Lumber

Charlie Foran remembered the wholesale trade, and one company in particular who preferred Canadian lumber:

> *We used to sell a lot to A.C. Crombie & Son in New York, and he would ship it to Argentina, South America. And they used to buy just twelve inches wide, no other width. They would buy 1 x 12, 1 1/2 x 12, and 2 x 12. And a man named Mr. Healy used to come. He was an old timer. I used to take him up to see the log drive, he'd love to see that log drive. I used to take him up to see that, and we got to be good friends. We had to make a stencil, and stamp, stencil every board "Canada" on it. Every board. We had a man stand right there with a stencil and lampblack and kerosene oil. We would get a stencil made over to Saco-Lowell. They had a gadget they would make them out of there; we'd stamp Canada. We'd get two or three made and this fellow would stand right there and stencil every board. Well, after awhile, I got acquainted with him, I said, "Mr. Healy, I've been wondering about something for a long while." "What's that Charlie?" "Why do you stamp Canada on those boards?" He laughed, "I'll tell you. The first order that we ever got that shipped to Argentina, we bought it in Canada. Then they wanted all their lumber to come from Canada. We bought all over the United States and we stenciled it Canada. They don't know the difference." They wanted it to come from Canada. (1407.066-067c)

> *[Deering] shipped it mostly to Massachusetts. He shipped carloads to New York, New Jersey. We shipped lumber to Argentina, we shipped lumber to Spain. But, it was through New York wholesalers, you understand. We loaded boats in Portland to go to Spain. We loaded on the cars to go to Argentina, they shipped that to New York. The wholesale outfit would buy carloads from Deering, and a carload somewhere else. They would get four or five carloads and put it on a dock in New York and ship it to Argentina. But that was a New York wholesaler.... We used to ship to a lot of retail lumber yards, but Deering would sell it mostly through a wholesaler. Like for instance, Newburyport, Mass., there was a

lumber dealer there. Deering didn't sell directly to him. Some wholesaler would buy it and they'd have it shipped there. They'd pay Deering and the retailer would pay them. (1407.022c)

What kind of wood was manufactured at the Deering mill? Charlie Foran put it this way:

*It was mostly one-inch boards, pine boards [*and one-and-a-quarter up to two-inch*]. We sawed some hemlock and a few spruce, whatever spruce would be mixed in [*both one-inch and two-inch*]. He sawed mostly white pine. Hemlock [*was*] used for building lumber. Nowadays they don't use much native hemlock, mostly western stuff. In those days we used a lot. Deering used to wholesale a lot, he furnished the retail department with pine boards, finished boards and matched boards, tongue and groove boards. He made some pine finish, but they bought a lot of their finish. We had southern pine and fir, inside house finish. It was cypress at first when I first worked there. They had North Carolina pine and fir, and then Deering used to make some white pine. There were some customers that wanted white pine. They'd make that special. (1407.021c)

This is how Ralph Morin and Charlie Foran remembered the wholesale trade and particularly the impact of World War II. In addition, Joseph Deering remembered clearly the War years:

*It was very difficult. You see, a tremendous amount of stuff was going overseas, and most of it had to be boxed. The crating lumber was hard to come by, and it was a tremendous shortage in lumber. A number of different projects were undertaken. (1402.054c)

J.G. Deering sawmill — early 1940's — showing boiler room, shaving collectors, scrap wood storage. Lumber was sawn at 2nd floor level moving from left to right. Third floor level was saw filing room. Man to left unidentified. Edward V. Corney to right. Courtesy J.G. Deering & Son.

Left to right — Albert E. Whitehurst, Charles Foran and Joseph G. Deering stand circa 1935 under the crane at green lumber brow, Deering sawmill, Biddeford, Maine. The saw blade was hand-forged in Philadelphia by B. Rawland in 1866 to be used in one of Deering's earliest up and down mills on Springs Island. Courtesy of Anne (Whitehurst) Ordway, Freeport, Maine.

Lumber Production for the War, 1940

The Deering mill had a reputation for being fast. A major "project" for Deering was running two shifts in 1940 to keep up with demand for lumber created by the war. It is worth keeping in mind that coincidentally this is the same year for which we have A.C. Cunningham's log book of the river drive. Charlie Foran tells the story this way:

*1940 I ran it two shifts. We sawed 7 1/2 million feet of logs, in five months. We sawed half of it for himself and half for the Diamond Match then, Diamond International now. Sawed half for them and half for us, see. The way we worked it, we'd saw a month for the Diamond Match. We put the loads in the field for them, piled it up the way they wanted it. Then we'd stop for them, clean everything off, and saw a month for J.G. Deering & Son. Wasn't Deering Lumber like it is now, it was J.G. Deering & Son. We'd saw a month for them. Each one could have some lumber drying on the stick. We did that in 1940. And I worked the two shifts. We only did it that year, darn near killed myself, to tell you the truth.

I worked from five o'clock in the morning until nine o'clock at night. Anyway, when we did that I had worked there quite awhile. Mr. Deering had asked me what I thought about two shifts. Well I says "Joe," I always called him Joe because he and I went to school together, when there wasn't other people around. I said, "Joe, I don't know anything about two shifts. All I know is what some of these fellows that works here [who] come from northern Maine, tell me. The second shift staves up what money the first shift does. They really wreck what you earn on the first shift." He said, "The Diamond wants me to saw for them." I said, "The hell with the Diamond! Look after you. Never mind them."

Well finally he said that we were going to saw for the Diamond too. I didn't like it, to tell you the truth, but I couldn't do anything about it. "Which shift do you want?" Well, I think I've worked here long enough that I ought to have my choice. I know the first shift is going to do more work, because we've got to look after the planing mill and the shipping. I'll take the first shift. He said, "All right, I'll get somebody to do the second shift. We're going to start at five o'clock in the morning. We're going to work seven hours a day, six days a week, each shift. Forty-two hours a week, each shift."

Deering says, "You know two-thirds of the people that come here for jobs,"—you know there's a certain type of people that come to a sawmill for a job and he says, "You know most of them. I want you to pick out the men for both shifts." I said "All right."

On Monday morning we started. We were supposed to start the second shift on Monday afternoon. When you overhaul a sawmill or any kind of machinery you've got to try it out a little bit, you've got some little things you've got to adjust. So I wasn't going to start the second shift on the first day when we have to stop three or four times to adjust this and adjust that. You had all power rolls that went by belts and steam. I was going to get those smoothed up, and start the second shift tomorrow.

He came around in the afternoon, wanted to know where's the second shift, I told him I didn't start it. "What's the matter?" Well I said, "Joe, we want to get the first shift running smooth, and then we'll start the second one. Start it tomorrow." He didn't like it, but, that's all right, too late. Couldn't do anything about it. Just scolded me a little bit, you know, and that was all right.

The next day we started the second shift. On the first shift we had old timers that were used to the machinery and the helpers. On the second shift for the machinery we had men who understood what to do, but the helpers, they weren't all experienced, you know. Some of those jobs were very important; you could hold up production like nobody's business. So a couple of places I put two helpers, to teach the new fellow how to work. By and by, I was out doing something else, and he [*Deering*] went up in the mill and he saw two helpers, he had two helpers there. He sent for me right off. Oh boy, he didn't like that. And I always called him Joe when there wasn't strangers around, I said, "Joe, these boys don't understand this job. We'll hold up the production mighty quick. Now let them go a day or so, and we'll take them off." "I'm not going to pay for two men on a job where there's only one..." Well I said, "You tell me you're going to get somebody else, go ahead and get them. I don't give a darn, one job's enough for me." I said, "I'll tell you what you'd better do. You keep the hell out of the sawmill. You stay in the office where you belong. I'll tell you when to come in here." So out he goes. I kept those couple of kids for a couple of days, until they got used to it. And I took them off, going fine.

Well anyway, the first shift with the old timers they could cut more in a day than the second. We had a surveyor, you know, for the morning shift, [*and*] one for the afternoon. There was a board to add it up, see how much we sawed. The second shift was, quite a lot under, you know. And I wasn't sure what to do about that. The boys would fuss at me, "You don't get nothing with the second shift. They don't do a thing." We were losing money. When we got done with the second shift the next day at nine o'clock at night, I went over to the surveyor at the board. They'd all crowd over to get him to add it up, how much it was. And I've forgotten, it was three thousand feet or something less than they did in the morning.

And I said, "Boys, I got something to say. You ought to be ashamed of yourselves. Why? They're all old duffers on the first shift. You've only got one old man on the second shift. Narcis King on that edger.... He's the only old timer here. The rest of you are all young. Old Tom [*Charron*] that runs the saw in the morning, he's 63 years old. They are sawing the pants right off you. You ought to be ashamed of yourself. What do think of me, going in to face Deering tomorrow morning? He's going to call me right in there. What do you think of me?"

The next day, so help me God as I sit in this chair, the next day they got more than the first shift. And the first shift never caught up to them after. Five months, never caught up to them after.

[*Were they new men?*] The men on the machines wasn't, the helpers were new. And never caught up to them after. We sawed more board production per hour than Deering's sawmill ever sawed before in its life. We sawed 4375 feet per hour for five months, average. (1407.009-014c)

Charlie's spontaneous, detailed recollection of running two shifts gives a sense of what a fine accomplishment it was for the J.G. Deering & Son sawmill. It was superhuman, yeoman service in a time of need. Lumber had to be produced for the war effort. Despite the combative discussions between the two men, Deering admired Charlie's work:

*Charlie Foran came in as foreman [*in 1933*] and during World War II he set up two crews—we ran a day and night shift—as a contribution to the war effort. Charlie ran both shifts. How he ever did it I don't know, but he did. He has a pacemaker now and he's not very well, but you still see him around the [*Saco Valley*] shopping center. I went to school with him. We both were in the grade school. He's one of two people who ever offered his life with me. [*During the 1936 flood*], the police came to take the crew away, did take most of them away, but Charlie and one other, [*Fred Hubert*] said, "No, we'll stay with Joe". This was during the flood. They thought that one of the dams was going to let go upriver and it would come down and rush and clean everything off the island where the plant is now. Of course it was a fantasy. It never could happen. (1402.050c)

The comment, "How he ever did it, I don't know", may appear at first to be faint praise, but for Joe Deering I would wager it was not. Charlie Foran also remembered the flood, and the other person who "offered his life" with Joe:

*Thirty-six, we had the flood. Across the street from Deering's there, where the tire shop is now, we had a half a million feet of lumber piled up there sawed special order for A.C. Crombie & Company, New York, piled up to dry. And the piles, sixty rows high one inch, two inch thirty rows high. The flood come in there and they'd lift up off of that blocking and slide out there to the river and they would go around and around in a whirlpool like this and all at once they would scatter just like a bunch of matches and go down the Atlantic Ocean. We couldn't do a thing about it. We put some booms around there and they snapped them just like nothing... I went to work on Monday morning and I never went home until next Monday night. Stayed right there day and night, Deering and I and another

An aerial view of Springs Island, Biddeford, Maine, and J.G. Deering & Son (front and center). Diamond Match Company (background) is across the Saco River. Note logs near the mills and the holding boom upriver, circa 1940. Site of first Deering Sawmill—1866—is at the dam (left foreground). Courtesy of J.G. Deering & Son.

fellow named Fred Hubert. Fred Hubert worked on the drives too. He worked on the drives and he used to bring our logs down [*from*] where we stored the logs up the river. He was the fellow that went up with the boat. He let down a section at a time, come down to the mill. He supplied us with logs, in other words. (1407.078-079c)

If one had to list the most frequently told stories by Foran, Deering, Blaney, and others, the 1936 flood, the 1938 hurricane, and the two shifts run in 1940 for the war effort rank high on the list. Another story told by Charlie Foran describes how important the men were to getting the job done. At every step of the way, the workers' skill, judgement, and hard work would make or break the Deering enterprise—from timber cruising, to river driving, to work in the sawmill—each person played a crucial role—even Conrad Dubois:

> "I'll tell you another one: the year we worked two shifts. Five o'clock in the morning is mighty hard to get everybody working. There was a boy named Conrad Dubois; he was one of the best workers, once he got to work, that you ever saw. And he worked in what we call the sticking hole. Taking the stickings and putting them in the box. And he was late every cock-eyed morning. I used to get so mad, I got a hot temper, you know. He lived on Main Street, Biddeford; you go down over the hill, next to the bridge, and he lived there. He and his father and his brother worked under me once in the woods, in the wintertime cutting wood. Wintertime I had to have crews cutting wood, and I'd go around, had Conrad cut, looked after him... Of course they worked piece-work, and I didn't give a darn if they worked or not. They only got paid for what they cut. But up in the mill, I needed Conrad every minute. And [*if*] he didn't show up, I'd go get him. Put another fellow in his place. I'd go in, and his mother would be up. She was a little stout, short woman. And I'd say, "Where's Conrad?" She says, "I'll wake him up." She'd go wake him up. Well this particular morning she said to me, "Charlie, come

with me. I'm getting darn sick of waking him up. You wake him up."

He was in bed. We shook him, "Come on there, get up and get to work, what's the matter with you." He got up and got to work. So, a few days afterward Conrad didn't come again. So I go down. I didn't bother with her. It was getting dark, like this time of year; well anyway I walked into this flat like this, go down the hallway, the room was on that side. Had no lights, but I knew where his room was. I felt my way in there, I got in there, I felt for somebody in bed, and I shook the daylights, I said, "Get up, damn you, get up. Time for you to go to work." It was her. 'Twas her, it wasn't him. She woke up and she says, "What the...." "I'm after Conrad." She says, "I'm waiting for him he hasn't come home yet." Did I feel cheap. [*Laughter*]. I never went after him again. I said, "Conrad you either get to work, or you pull out." (1407.032-033c)

Portable Sawmills and the End of Lumber Manufacturing in Biddeford

The final chapter in the story of J.G. Deering & Son's lumber manufacturing in Biddeford brings us to the use of portable sawmills in the Saco River Valley. When asked how the Biddeford mill kept operating after the last river drive in 1943, Charlie Foran responded:

*Mostly portable mills, and, I forgot the last year that we cut there [*at a portable mill site in Sweden/Lovell*]. And we cut a lot of logs local, then we dumped them in the fields up here near the brook and we floated them into the river in the spring. (1407.073c)

But the portable mills were not able to supply the Biddeford planing mill with the quantities needed to match production of the river-drive years. Lumber would come in by truck from portable mill sites in Lovell, Sweden, and Fryeburg. Charlie Foran remembers Ralston Bennett trucking lumber:

*He used to bring it in from the portable mill, sawed lumber. After it was sawed, we used to handle it and dry it and one thing or another. [*Would it be planed?*] Oh yes, we had to stick it up and dry it.... Cunningham's boy had a truck, oh there were three or four others that had trucks who used to bring stuff in.... We had to [*buy*] some [*lumber*] besides that. They couldn't keep us going. When we had the [*Biddeford sawmill*] going, Deering had three or four portable mills going all the time, besides the mill.

Yeah, going about all the time. Different areas, each mill would saw probably a million a year, something like that. That isn't the exact figure but something like that. Three or four of those mills going most all the time. (1407.075c)

Although three or four portable mills might produce three to four million board feet, that still is perhaps half of what the Deering mill produced—its record being the 1940 production of 7 1/2 million with two shifts. On the other hand, if one reviews the summary of board footage produced from "upcountry logs" in any other year, the figures are comparable to what the portable mills were getting out of timber lots. But the portables took a big chunk of the work away from the Biddeford mill, that is, squaring timber and re-sawing into boards. All that was left to do at Biddeford was drying, planing and sales. As we have noted in earlier chapters, many other forces conspired to gradually bring an end to lumber manufacturing in Biddeford. In addition to the shift from a war-time to a peace-time economy, J.G. Deering & Son began converting to a predominantly retail lumber business. Phyllis Deschambeault worked in the Biddeford office for Joseph Deering (with Arlene Chappell and Irene Maher). With time out from 1937 to 1945 to begin raising a family, she worked for Deering from 1929 until 1958 when she was elected York County Registrar of Deeds. Irene Maher said Phyllis knew the lumber business so well that she "must have had sawdust in her blood." Deschambeault remembered considering the wholesale trade "spasmodic," while the retail trade became increasingly the "mainstay" of the company, especially in the 1950's. (2194.004c) Ralston Bennett describes the change from his perspective in Deering operations in the 1940s and 1950s. I asked him if there were portable mills around when the drive was still going in the early 1940's:

*I can't remember that there was any. No, I think they drove everything. I can't remember that they started too much before, well, we had one sawmill over here to Sweden. In wartime, that's the first time I ever remember anything about a sawmill, we sawed for the railroad ties and things like that, for the shipyard [*in South Portland*], because I hauled those. That's when [*Erland*] Babe Day was foreman, and Mr. Cunningham. They were in together. Mr. Cunningham was over everyone, next to Mr. Deering. Babe Day was a woods foreman. Leo was usually scaling on the river....

[*Were there more portable mills after the war?*] Seemed to be, yes. Yes, that's just about the time [*1944-1945*] they started to [*use portable mills more*] then they had the mill set up over here in Lovell once, in a field. That was in wartime and they were bringing stuff in there to it. I was hauling from that yard. Also, I was hauling logs. There were quite a few different guys hauling out of this mill at that time. There was a fellow out of Cornish he was hauling some for us.... Carroll Perkins, that's the guy. He drove the bus too, for the help.... What they would do is bring the bus load up and then some of them would go into the woods....

[*Would you rather haul logs or sawn lumber?*] Well, I don't know, didn't make too much difference. When I drove myself, the truck I drove myself,

I would rather haul the logs. But some of the drivers would rather haul the lumber. Wasn't so much lifting to it. (1423.010- 011c)

Bennett and the other truckers were paid by the thousand board feet for hauling logs. When I said to Ralston Bennett that I sure would like to talk to someone who ran a portable mill, a sawyer maybe, he said:

*I know where there is one, if you can get a hold of him, Chet Leonard. You can't beat him. He was the one that ran the mill while we was in Sweden. He was foreman for the mill, he ran the mill and sawed himself. (1423.013c)

Chet Leonard was a sawyer and portable mill foreman, at various times, for Diamond Match, J.G. Deering & Son, and the Ellery Clark mill in Hollis Center. His father was a sawyer, and Chet began working in New Hampshire mills in 1933. Although Chet worked for J.G. Deering & Son for just six months in the early 1950's, his description of how a portable sawmill worked gives us a sense of what it meant to take the sawmill into the timberlot.

*I operated a sawmill, the lumber operation and the sawmill for Deering Lumber Company. That was in Lovell, Lots Four and Five. The lots in Lovell go by number... As I recall, this particular mill was all set up, had been operating. There had been some dissatisfaction about getting its production. Babe Day, he contacted me first and told me about it. Previous to that I was working for Clark Lumber Company in Hollis Center. They sort of had a little tough luck, they folded up after the fire [*1948 forest fires in southern Maine*], and that left me without a job, so I was job-hunting.... At the time I worked for Clark Lumber Company, there was not too many lumber operations going on around here. Soon after that, they started coming in pretty fast. There was a lot of timber in this area and they started cutting it off pretty fast. I can't remember the names, there was two or three other mills around. There was a Lewis Lumber Company, down at Portland I think it was. Owned by Bennie Lewis, he operated a mill over here, for a few years he was in business here, around that time when I was working for Clark. As I say, Biddeford hired out and this Jack Trot [*from Deering said*] take over the mill in Lovell, which was down at that time. I stayed until, that was in spring, I stayed until fall. I had an opportunity, Diamond at that time was getting ready to come in to Fryeburg. Diamond had lots of operations in Fryeburg as far as buying timber, timber lots. Lawrence Gray was their head man, head forester, and he bought up a lot of timber land. So he recommended me to Mr. Feeny, who was manager of Diamond National, Diamond Match at that time. Donald Feeny.

So I talked with Mr. Feeny, I went to him for an interview, about getting into the finishing mill down here in Fryeburg. Because it was a stationary job. I think we moved, my wife and I moved around nineteen times chasing these sawmills around, all the way from New Hampshire to Vermont to Maine. Had no roots anywhere, no more than get acquainted and then pick up and move again. I liked the sound of what Mr. Feeny had to offer, so I took the job. (1426.004-006c)

Chet Leonard remembered clearly the set-up at the Deering portable mill in Lovell:

*Well, in the mill itself it took one, two, three, four, four to five men to run the mill. Then we had a tractor driver, two tractor drivers. A scoot loader, that would be about eight. Then we had four to six men at different times cutting the timber in the woods. That was under my supervision too, cutting the timber.... (1426.008c)

*It had a head rig, an edger, and a jump saw. That's all there was. Run by a gas power unit. [*A jump saw is*] just a, a stick comes up through the floor and you put your foot on it and you push like this. It jumps the saw up through between the rolls to trim the lumber. If you've got a tag end that's going to bother, or waste, you don't want to bring it out with your other lumber. It's just waste piece, you just rim that off, before it goes through the edger or before it goes by on the rolls out to the pit.

You had a log turner. We used to call him the roller, and then sawyer, and then the man behind the saw taking away from the saw and running the edger, he did the two jobs. He ran the jump saw and the edger and took away. There was another man that I left out. I said five men in the mill, uh, five to six, because you had to have a man behind the edger to take care of the edgings and throw the slabs across the track. After I had left Deering Lumber Company, they put in what they called a slab kicker. I never saw one operate, but it's a large drum with spikes in it that turns pretty fast. Drop the slabs on there and it would shoot them up into the air, and they could angle it for heights in any direction they wanted to. Shoot the slabs right out into a heap. Before that, they had to have a man in the pit. I am coming up with another man. Another man in the slab pit hauling the slabs out. If he didn't, it accumulates in there in a short time. (1426.010-011c)

This particular mill was powered by a gasoline engine, but others were diesel or even steam powered, with the boiler fueled by sawdust and slabs:

*Well, with steam you had a fluctuation of power. It depends whether in the wintertime the slabs were all ice. Sometimes the fireman would have a little difficulty keeping up steam. If you didn't have the steam you didn't have the power. Of

White Pine on the Saco River

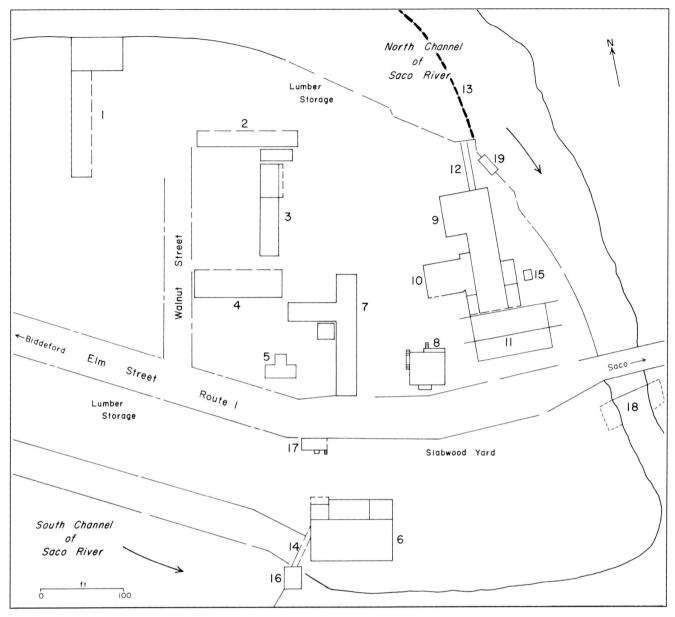

J.G. Deering & Son, Biddeford, Maine. Plan of lumberyard and manufacturing facilities on Spring's Island approximately as in the 1940's.

Key
1. stable (top) and wagon shed
2. shingle shed
3. lumber
4. lumber and molding
5. Moses Bradbury House (1795)
6. garage and storage
7. doors-windows (top), finish pine
8. office of J.G. Deering and Son (formerly Springs Tavern 1798)
9. planing mill at ground level, sawmill 2nd level (1913)
10. boiler and engine rooms
11. green lumber brow platform and overhead crane
12. log chain
13. log boom
14. site of late 18th C. water-powered sawmill (Bradbury), in 1866 became site of the first Deering sawmill
15. waste hog
16. CMP Co. dam
17. nails
18. sites of early water power sawmills operated by Deering in the late 19th and early 20th C. Now a CMP dam
19. boom house and privy

course with gas or diesel you had a certain amount of power available at all times. Maybe it wasn't enough, but you knew what you had. (1426.012c)

What sort of wood did Chet cut for Deering?

*We supplied his, or helped supply his retail business in Biddeford—pine, hemlock and spruce. And then he cut some good hardwood, especially yellow birch or rock maple. The good quality, he used that for the door stoops.... Your hemlock was mostly dimension, spruce and hemlock. You made as few boards as you could. The boards were not so much in demand. We put it on rolls, kicked the trig on the rolls, rolled right on to Bennett's truck. He would haul them into Biddeford.

The hemlock and spruce would probably go as soon as possible, if you could get to it, to the finishing mill. The pine would be stuck up and dried.... They used to think they had to dry it a year, but as time went on, six months was good enough. (1426.015-016c)

Deering had some substantial timber lots in Lovell, Sweden, Waldoboro, Winthrop, and elsewhere. Leo Bell remembers how expensive it was to move a mill:

*I think setting up a mill, taking it down, setting it up again on some other lot, I think we used to reckon around $1500. Of course, that was thirty years ago. It would be three times that now. (1404.024c)

Leo remembered portable mills staying on these large sites for one or two years.

*Winthrop, we had a mill there. That mill was there a year and a half, two years. And we had mills in Sebago, all around. There was one on this little mountain [*Mount Tom, in Fryeburg*] right over here [*points out kitchen window*] that was there for year and a half, two years, 1955–56. (1404.027c)

Deering shut down the Biddeford sawmill in 1948, continued with portable mills and began to buy lumber wholesale from other manufacturers. With the Saco River drive over for good in 1943, and the Deering Biddeford sawmill quiet, clearly an era had passed. One of the themes in this book has been how much the success of the operation depended upon the energies and talents of individuals. Without Leo Bell, A.C. Cunningham, Charlie Foran, Ed Burrill, Ken Blaney, and dozens of others—this lumbering enterprise would have been quite different. By and large, these people spent their lives in the Saco River Valley, and although we have spent much of this book talking about how exceptional their work was, they will tell you that they were just making a living.

Conclusion

The history of lumbering on the Saco River has not been adequately told. This book provides only a part of the story. We have explored, primarily through oral sources, the business of operating woodlots, driving logs on the Saco to Biddeford, and manufacturing lumber at the J.G. Deering & Son mills. We have examined in detail only the period from 1920 to 1960, with primary attention given to the Deering operations. The history of the Diamond Match Company on the Saco is an untold story. We know little about the Androscoggin Pulp Company's operation at Steep Falls—except that at one time it also drove its wood on the Saco River. We have only touched upon lumbering activities in the eighteenth and nineteenth centuries. However, we have learned a great deal about Deering's activities on the Saco in the twentieth century.

Deering's Saco operations were perhaps one-twentieth the size of the massive Penobscot River driving operations in Argyle—in terms of timber cut and the number of people employed. Our study has focused on the lives of a few people in the Saco Valley and their ability to earn a living in this part of northern New England. Unlike the Penobscot, the Saco winds its way through relatively settled agricultural areas. Woods work and river driving were just one part of a living made from the land. Leo Bell tended his corn, managed a corn packing facility, cruised timber lots, and drove the river only to Hiram—so that he might return easily to his Fryeburg farm. This was not a river where the crew stayed together in camps for months on end. Boarding houses were convenient, but many of the men were able to return home, at least every few days. We have explored how woods and river operations in the well-settled Saco Valley relied on many long-term, trusted people—Frank Brown, A.C. Cunningham, Charlie Foran, and Leo Bell. J.G. Deering & Son employed workers as they were needed from a large pool of independents-farmers, woodsmen and craftsmen. Real labor shortages for Deering began in 1940 with the onset of the war.

In addition to learning how timber was procured, driven, and manufactured, we have seen how a family-owned business kept at this complicated and labor intensive work. With this series of biographies, we have also seen how heavily Joseph Deering relied on Cunningham, Bell, and Foran in particular, and how those men organized many others to keep the show on the road. The reader will recall that Conrad Dubois was equally important to Charlie Foran as any other member of his mill crew.

Finally, with the memories of the people interviewed for this book, we have documented the end of a three hundred year span of lumber manufacturing directly on the Saco River. The history of J.G. Deering & Son in the twentieth century could not have been gathered without the personal stories told here. Another book should be written based upon business records and other documents that would take the story back to the eighteenth century. Such a book could also broaden the inquiry to other companies and other people. *White Pine on the Saco River* is primarily about J.G. Deering & Son and the people who worked for that company. These people are the best sources of all—they were there.

Afterword

by Thomas M. Armstrong

It may be of interest to the reader to know the progression of the Deering Company following the end of Michael Chaney's story—with brief explanation as to how this narrative came to be written.

My wife and I moved to the Biddeford-Saco area in 1955, and after three years with J.G. Deering & Son, I purchased assets of the Company's building supply business in 1958 and the Biddeford lumber yard in 1960. Three generations of Deerings had owned and operated from this site for ninety-two years. Joseph Deering was to retire, closing out the remaining portable sawmill, the machinery and equipment, the unfinished lumber inventory and 3500 acres of timberland.

No longer manufacturers of native lumber, we concentrated on modernizing warehouse facilities and expanding lines of building materials to meet the growing needs of contractors, homeowner and industrial customers in York and Cumberland Counties. Acquiring adjacent property on Springs Island in the 1970's allowed the construction in 1980 of more adequate display and retail store facilities.

One of our sons, C.D. Armstrong, assumed management of the Company in 1987 as its fifth owner, on the same site, over a 121 year span. He has already expanded his business with a second location in Kennebunk.

During all this time I had maintained a personal interest in forestry and an increasing awareness of the Deering Company's role in the Saco Valley lumber heritage. It was time now to record a small portion of the Company's history and to acknowledge more adequately something of the three hundred years of continuous lumber production on the Saco River. The day of the river drive was over. Hundreds of mills had come and gone. J.G. Deering & Son was, to our knowledge, the last sawmill located directly on the Saco.

The larger timber operations on the Androscoggin, Kennebec and Penobscot Rivers in the nineteenth and twentieth centuries are well documented. Bangor was the lumber capital of the world; the colorful "Bangor Tiger" river driver/logger captured the imaginations of many in song and written word. In contrast, the Saco, though smaller, has an earlier and distinctive history of its own. From the arrival of its first settlers, timber has been of paramount importance to the economy and the lifestyles of its inhabitants. But there is today very little of the collected or written history of this industry in the Saco Valley.

By good fortune I learned of Sandy Ives at the University of Maine and his work in recording, for the future, the lives and times of people past and present. He agreed to co-sponsor a project in 1980 to capture the knowledge and recollections of men associated with the log drive and sawmill operations of Deering and other companies. Michael Chaney, a former student of Professor Ives, completed this work and the tapes and transcripts were filed at the Northeast Archives of Folklore and Oral History in Orono.

Chaney went on to pursue his career in other areas and it was not until 1989 that we seriously discussed weaving the transcripts into a published story backed by photographs and additional information. I am greatly indebted to Mike for completing this narrative while serving as Resource Center Director of the New Hampshire Humanities Council and, with his wife Laura, raising three sons and a daughter. A task of no small accomplishment—this required considerable patience, commitment and "overtime!"

Additionally, I must express deep personal thanks to Professor Ives as Editor and Publisher, to Pauleena MacDougall as Managing Editor, and the Northeast Folklore Society and to the University of Maine for the printing of *White Pine on the Saco River*.

There is no longer the clank of a log chain nor the whine of a saw on the Saco. The nature of this business has changed. Yet quality white pine, hemlock, hardwoods and smaller stands of spruce continue to grow abundantly in the Valley. Highly mechanized loggers carry on the harvest of this timber on a fifty to seventy-five year rotation.

Strategically located modern sawmills in New Hampshire and Maine produce lumber from these logs carefully graded to industry standards. And it is to a number of these mills and other forest product companies and organizations (as listed in the front of the book) that we tip our hats for generously assisting in the costs of printing and publishing.

The Forest Products industry continues as a strong and vital element in the Valley's economy. It will remain so if sufficient productive timber land is maintained for managed forestry purposes by both small and larger landowners.

Frank C. Deering's advice to Maine farmers in 1901, when he estimated the cut along the Saco for that year to be thirty million board feet, still carries a relevant message:

> —Keep your pine growing. Keep it ahead of you all the time. Look carefully after the growth you already have....
>
> Two hundred and fifty years of an honorable industry is a record that should appeal to every farmer and mill owner on Saco River and its tributaries—it is (my) hope that the dawn of this glorious [*twentieth*] century may also be the dawn of another two hundred and fifty years for this lumber producing valley.

(Printed in entirety Appendix No. Four)

APPENDIX ONE
Daily Log Book—Saco River Driving Co.
Kept by Asa C. Cunningham, *Woods and Drive Boss*
1940 Log Drive—Saco River and Tributaries
April 8 to August 13, 1940

1940
April the 8th we rolled in the greater part of the Little Ossippi logs. But they were badly frozen in the ice so we discontinued for a few days. The drive did not start until about the 23rd. 21st and 22nd we rolled in the balance of the logs at Edgecomb's Bridge and did more or less preliminary work.

April 21, 1940
Driving season about one month late. Snow has not begun to melt and logs are frozen in the landings.

April 22 and 23
Snowing—not working. We have rolled in the greater part of the Little Ossippi logs. Leo Bell driving Cold River with five men.

April 24
Merle and crew driving the Little Ossippi. Saw Mr. Hamilton regarding his booms. [*Cumberland County Power and Light.*]

April 25
Driving Little Ossippi—Merle and crew. Rear into the jam at noon. Men worked in afternoon on Steep Falls holding boom.

April 26
Started to roll in the Great Ossippi logs at Cornish. Was to Kezar Lake today. The ice looks strong. I think it will be a week before we can move those logs. We hung Hiram Boom.

April 27
Rolling in the logs at Kezar Falls. About fifty men and two teams on both landings. Water good—landings frozen—slow rolling.

April 28
Finished landings at Porter Bridge. Started the rear at Porter Bridge at noon today. Rear above Kezar Falls at night. J.G.D. was up taking pictures of the roll ways. Leo Bell came down to tell me we could move the Charles Pond logs tomorrow.

April 29
Driving Great Ossippi. Rear at the Power Dam tonight. Six men working. Leo Bell and crew at Kezar Lake. Our motor went on the bum and we did not accomplish much. Still rolling landings on the Great Ossippi. Logs are still frozen. We have to cut them out of the ice. Took the motor over to Lewis tonight.

April 30
Rolling in landings on the Great Ossippi. Rear tonight at the Pendexter Landing. Leo Bell and five men on Kezar Lake logs.

May 1, 1940
Rolling in landings on the Great Ossippi. Rear at Hell Cat Eddy tonight. Water came up four feet in the canal. Boom out of Kezar Lake. Water too high to turn over Brownfield.

May 2
Rolling in Great Ossippi logs. Rear still at Hell Cat Eddy. No motor and we could not do much with one boat. Put lines on Steep Falls Boom. Saw Mr. Hamilton. He says his booms will be ready to put out in a week. Water too high to put on boards. Kezar logs out of the lake. Water has come up six feet in the canal since we started.

May 3
Finishing rolling in the Great Ossippi logs. They were still frozen. We have to chop them out of the ice. Finished Hiram Boom. Will have to put in another boom before it will be safe to hold logs. Leo Bell and two men working on the Kezar Lake logs. Rear at Spring Bend tonight. Not working on Ossippi logs as water is still rising. We put more lines on the Steep Falls Boom today.

May 4
Started the rear on the Great Ossippi at Hell Cat Eddy—six men working. Five men and motor boat on Pleasant Pond logs. We got the booms of the Meadows. Water came up about two feet in the Pond in the last day or so. Steep Falls Boom holding good—four men putting on lines to keep the logs from going under the boom. There are some big cakes of ice and drift wood ahead of the logs. Rear of the Ossippi to Cornish Flats.

May 5
Went to Steep Falls Boom six a.m. Boom broke away from Limington shore. One of the rings parted in the weld. Worked all day with 12 men getting it back and putting on more lines. Lost about 25,000. J.G.D. and Carney were up. Water rose about two feet. Snow seems to be melting fast in the mountains. Bell's crew not working today. Did not work on rear.

May 6
Started the rear of the Ossippi at Cornish Flats—at Highland Rips tonight. Three men driving Charles River—rear out—four men and a team getting boom for Hiram. Water dropped about 4 inches in the Main River. It will be 10 days before we can turn over Brownfield anyway. A lot of snow yet in the mountains. Rear at Gould's Island.

May 7
Finished getting out the Hiram Boom. Water has dropped at Hiram about 18 inches from the high pitch. There is still three feet more to go before we can turn over Brownfield. No men working at Fryeburg today. Holding booms all o.k. Rear at Half Moon Pond. Came down the river the C.C. [*Cumberland County Power and Light*] have not done much at their booms yet. Water seems high at all the dams.

May 8
No men working at Fryeburg. Rear of the Great Ossippi Drive in to Steep Falls Boom. Hauled the boats back—two men started on the Little Ossippi. Water dropping but still too high to hang our new boom at Hiram.

May 9
Started to run our Little Ossippi logs over Chase Dam with 14 men. Working on the Kezar Lake logs. The water has dropped about 3 feet at Fryeburg. Some of the Wakefield logs got away on the High Water. We will have to pick them up.

May 10
Turning Little Ossippi logs out. Bell and five men working on the Kezar Lake logs.

May 11
Finished the Little Ossippi. Had to leave 50,000 above the dam. There was not room enough to get them over. I had to clean it by Chase's Mill so as not to block him from getting logs to his slip. Two men booming Bonny Eagle: Robert Durban, Arthur Tuffs. Water dropping in the Main River—two men driving Charles River.

May 12
Swung the new boom at Hiram and started to turn in the Wakefield logs. Some of them scattered in the Maples and Dragon Meadow. Two men booming Bonny Eagle Pond:
    Robert Durban
    Arthur Tuffs

May 13
Working at the Wakefield logs. Some high and dry due to high water. Two men working at the Harbor on the rear. Two men booming Bonny Eagle:
    Stanley Harnett
    Harold Burrill.
Came from Fryeburg to Hiram in a boat today. Water is high but dropping will be o.k. in a few days. Saw Mr. Hamilton. He says his boom at Bonny Eagle is ready to put out but would be afraid to turn logs in now with the boards not up.

May 14
Working at the Wakefield logs with 18 men. High water carried about 50,000 into the river. Will have to pick them up. Two men on the rear at Old River. Water dropping—looks like rain—will be running over Brownfield in a day or so if we don't get a rise. Let the Fidelity Landing to Phil Hartford to roll in. Saw Mr. Hamilton—he says he will hang his Bonny Eagle boom tomorrow.

May 15
Six men on the Old River rear. Rear into the body at night. Merle and crew working on Brownfield. Some of our logs from Pleasant Pond due to the high water came over the bank. Merle and crew picking them up. Water dropped about 8 inches since yesterday.

May 16
Ten men picking up logs on Main River on Brownfield. Five men working on Limington boom. Will take another day to complete it.

May 17
Started to turn over Brownfield from Old River—8 men working— river bank full and current good. Six men and boat booming Bonny Eagle Pond:
    S. Harnett
    Edwin Foster
    H. Burrill
    Ed Burrill
    William Wintell
    Ivan Warden
Two men cutting braces and pick polls for Limington boom.
The C.C. Boom not out at Bonny Eagle. Looks like rain and there is a lot of snow in the mountains.

May 18
Water came up two feet in the canal last night and

kicked back into the Kezars. We did not accomplish much on our rear. Water began to drop in the afternoon at Swan's Falls. We just got the landings cleared today. Three men and a boat booming Bonny Eagle Pond:

    S. Harnett
    Ivan Warden
    Harold Burrill

Saw Mr. Hamilton today. He says that he will put out Bonny Eagle boom tomorrow.

May 19

Driving Old River—water changed—bank full and current running out—running over the banks in some places. I'm afraid it will be bad on Brownfield. Mr. Hamilton told me that he would finish the Bonny Eagle boom today. J.G.D. was up to Old River. C.C. boom not out—rear of Old River drive at the Old Barn. Put some more lines on Hiram boom—water is high and logs are piled in—will be hard to pick out. Three men and a team putting in the beaver Pond logs at Walker's Bridge.

May 20

Driving Old River. Two men on the Little Ossippi rear. Cleaning the dam so Chase can put on his boards. The Company did not have their Hiram boom out today at noon. Hauled the fins for Limington boom. Sent the boat to Pleasant Pond with three men to tow the boom across. They did not accomplish anything—head wind—three men and team turning in Beaver Pond logs. Rear of the drive at the mouth of Old River.

May 21

Seven men on the rear today. Four men and boat on Pleasant Falls logs—four men changing lines at Steep Falls and Hiram. Three men and team on the Beaver Pond logs. Was to Hiram Dam. The Company boom looks to me to be a very poor one—too small—and I think that we will have a lot of trouble getting logs through that narrow sluiceway. Hamilton and Bolter were there. Rear half a mile above the Bridgton Road. Motor went bad did not accomplish anything at Pleasant Pond.

May 22

Four men on the Pleasant Pond logs. Had motor repaired last night but could not do anything with it—same trouble—finished turning in the Beaver Pond logs. Opened Hiram boom—ten a.m. Notified J.G.D. office—Hamilton, Bolter, Harris of the C.C. were there. Carney came up—run through about 4,000 pieces with four men—works bad:

    Robert Dearborn
    Arthur Tuffs
    Stanley Harnett
    Erland Day

Falls below the dam seemed to work good—rear at the Islands tonight.

May 23

Putting fins on Limington boom with four men. Seven men at Hiram sluicing:

    Erland Day
    Kendrick Sawyer
    Stanley Harnett
    Laurence Eastman
    Robert Dearborn
    Arthur Tuffs
    Raymond Day

Boom broke and went over the dam at noon. It will take two weeks more to get the logs over. I am spending about all of my time at Hiram. It is slow. If we had had a proper sluiceway about 17 feet wide and a good boom I don't think we would have had any trouble. Russell Chisholm was drowned today below the mouth of Pleasant Pond. Drivers looked in the afternoon for the body. Sent two men to start logs out of Little Ossippi. Did not do anything on the rear—everybody searching the river.

May 24

All looking for the body. No luck—head wind on Pleasant Pond. We had ten boats and canoes dragging the river down to Brownfield. State Office, Gordon, two Game Wardens, Vern Black, and another one. Did not do any driving today. The C.C. getting out another boom from Hiram.

May 25

Four men on Pleasant Pond boom—got one over last night—most of crew looking for the body. Head wind on the Pond today. Motor on the bum. A.C.C. went up at night to tow on the boom. Head wind all night. Boat caught on fire and we had to bail out. Still head wind.

May 26

Had motor repaired—opened the boom at the mouth of Little Ossippi—head wind but some logs went away. Gave up looking for the body and started rolling rear below Pleasant Pond. Company boom not out yet at Hiram.

May 27

Ten men on Pleasant Pond—wind fair—we towed two booms across. Four men at Limington boom. Two men taking off lines at Steep Falls boom. No boards on yet at West Buxton on Bonny Eagle. Logs at Bonny Eagle to sluice.

May 28

Driving from Pleasant Pond to Big Eddy—started to turn over Hiram today at noon with five men:

    Erland Day
    K. Sawyer
    Will Foster
    Walter Foster
    Raymond Day

Water has dropped about 18 inches—logs run over better but pick out of the boom slow—three men at the Little Ossippi running logs out.

May 29

Turning over Hiram—worked fairly well—water rising—ten men:

    Erland Day
    K. Sawyer
    Will Foster
    Walter Foster
    Kenneth Blaney
    Raymond Day
    Ernest Foster
    Robert Dearborn
    Arthur Tuffs
    Harold Burrill

J.G.D. notified me to clean the Cornish Bridge of logs. We were a day and a half cleaning them off. When the logs began to run they loaded right up again. We have two or three jams a day that takes an hour to six hours to pick out Hiram sluice. The rear was at Lovells Pond.

May 30

Opened Steep Falls boom with four men. Started running over Steep Falls. Six men at Hiram:

    Erland Day
    Raymond Day
    K. Sawyer
    William Foster
    Ernest Foster
    Walter Foster

Water rose eight inches—two jams at sluice today—12 men on the rear. It's at Shepard River tonight. Logs out on the meadow.

May 31

Six men at Hiram:

    E. Day
    R. Day
    K. Sawyer
    E. Foster
    William Foster
    Walter Foster

Carney wanted the boards changed on the dam. Did not work as well as before so we had to change back. Four men at Steep Falls turning over. Two men and boat booming Bonny Eagle:

    S. Harnett
    R. Dearborn

Wanted to sluice over Bonny Eagle but not water enough—no boards on—don't know why.

June 1, 1940

Six men turning over Hiram:

    E. Day
    K. Sawyer
    E.A. Foster
    Wm. Foster
    Walter Foster
    Raymond Day

Boom let go at Steep Falls. Not hitched right. Sluicing at Bonny Eagle with 3 men:

    Sam Berry
    Ed Foster
    Frank Swinington

Water came up 8 inches at Hiram. Rear at lower Brownfield. Five men turning over Steep Falls:

    Harold Day
    Bob Durban
    Harry Taylor
    E. Day
    A. Tuffs

June 2

Brought the five men from Steep Falls to Brownfield on the rear. Can't turn over any more until company repairs their boom. Mr. Hamilton told me that he would put it back properly tomorrow. They did not have the right kind of hitch on it. Rear over Brownfield Flats. Six men turning over Hiram:

    E. Day
    K. Sawyer
    E.A. Foster
    Wm. Foster
    Walter Foster
    Raymond Day

Water rising—came up 20 inches at Swan's Falls.

June 3

Eight men at Hiram:

    E. Day
    R. Day
    K. Sawyer
    R. Durban
    W. Foster
    E.A. Foster
    Fred Hubart
    Bram Martell

We had four jams today at the sluice. The boom went over the pier. Ten men with M.C.C. on the rear. It was at Town Farm landing tonight. The C.C. put the Steep Falls boom back. D.A. Bradley was down to Hiram.

June 4
Had Hiram crew at Cornish Bridge today. Logs worked hard. Did not run any over Hiram Falls. Started to run over Steep Falls. Logs jammed on the guy wires and someone ordered them cut. Logs are going everywhere over the dam. I don't like it. They are picking up on the rocks below and going in the eddies. Started to sluice at 1 p.m. at West Buxton: Same Berry, Ed Foster and Frank Swinington. Logs working good. I don't think that they are damaging any at the sluice. Rear at Moose Brook tonight.

June 5
At Hiram sluicing Erland Day and K. Sawyer worked until noon on a jam at the sluice. Eleven men in the afternoon turning over: E. Day, K. Sawyer, S. Harnett, Fred Hubart, Bram Martell, E.A. Foster, Walter Foster, R. Durban, Arthur Tuffs, Harold Burrill and Ed Burrill. The C.C. put their boom back over the pier and spiked a timber on top of pier to hold it. Three men sluicing at Bonny Eagle and West Buxton: Sam Berry, E. Foster and Frank Swinington. Logs worked good. Nine men at Cornish Bridge on the pier jam. Logs work hard due to high water. Did not turn any over Steep Falls today. Water about the same. Saw Mr. Hamilton—he said Harris was coming tomorrow to see about putting the boom back at Steep Falls. He told me that it was Carney ordered the boom cut. The rear on Dragon Meadow at night. Mr. Scribner was down today looking over the logs and conditions at Hiram.

June 6
Logs jammed at Bar Mills. The C.C. boom was not hitched on the pier. We had to wait until they got a crew there to hitch it. Three men turning over at Bar Mills: Fred Hubart, Bram Martell, and Frank Swinington. We log 1/3 of a day with three men waiting for them to hitch their boom. Six men at Cornish Bridge: Erland Day, Robert Durban, K. Sawyer, Stanley Harnett, Wm. Foster and Harry Taylor. Swam Foster's horses on to Dragon Meadow. Had the rest of the crew on the dry logs. We got clean at night.

June 7
Had 20 men at Cornish Bridge today. 500,000 in a jam. Got the jam out at noon. E. Day, K. Sawyer, R. Day, M.E. Cunningham, Wm. Foster, E.A. Foster, Walter Foster, Bram Martell, Fred Hubart, R. Durban, Arthur Tuffs, Stanley Harnett, Harold Burrill, Ed Burrill, Emel Bass, R. Gilpatrick, Edmund Christolm, K. Smith, Edwin Foster and Frank Swinington. One half a day each at Cornish Bridge. Same men on half a day cleaning out Hiland Rips Dryway. No men working on the rear.

June 8
Turning over Hiram. Six men: E. Day, R. Day, K. Sawyer, E.A. Foster, Wm. Foster, and Walter Foster. Three men sluicing at West Buxton and Bonny Eagle: Frank Swinington, Ed Foster and Sam Berry. Had three jams at the sluice at Hiram. Rest of the crew on the rear. It was at Hiram Bridge at night.

June 9
Six men at Hiram turning over: E. Day, R. Day, K. Sawyer, R. Durban, Kenneth Smith and Ed Burrill. Three men sluicing at West Buxton and Bonny Eagle: Sam Berry, Ed Foster and Frank Swinington. Rest of the crew on the rear. It was at the Hiram Boom at night.

June 10
Two men turning over Hiram Dam: Erland Day and K. Sawyer. Rear over at three p.m. and started below the dam. Four or five of the men did not show up. I guess they were afraid of the Falls so I ordered some from Portland and Bangor. Water dropping. We started turning over Steep Falls. The Company boom works good. Mr. Hamilton says that they will put the boards on Bonny Eagle tomorrow.

June 11
Rear about half over Hiram at night. Run all the body of the logs out of Steep Falls boom today. Seven men working. The C.C. put some of the boards on at Bonny Eagle today. We had 3 men at Bonny Eagle and West Buxton sluicing: Sam Berry, Ed Foster and Frank Swinington. Logs working good.

June 12
Rear was below the falls at 2 p.m. The big Eddey below the falls is full of logs. We sluiced at West Buxton with 4 men: Bob Durban, Arthur Tuffs, Swinington and Ed Foster. About 1 million and 1/2 of logs in the pond. We had a big jam in Limington and a lot ran in the dry way. Did not sluice any at West Buxton. The rest of the crew working in the Eddey below Hiram.

June 13
Motor went wrong. Did not get away from the Eddey until 3 p.m. Rear at the Wife Rips tonight. Four men cleaning Limington Dry Way: E. Day, Ed Burrill, Stanley Harnett and Kenneth Smith. Four men sluicing at Bonny Eagle: Sam Berry, Frank Swinington, Ed Foster and Walter Foster. Three men at West Buxton: Arthur Tuffs, Robert Durban and Harry Taylor. The logs would not run by the Company's boom at Bonny Eagle after the boards were put on. We had to keep two men there to shove them by: Ed Foster and Walter Foster.

June 14
Got along good with the rear today. It was at the Cottages below Cornish Bridge tonight. Three men at West Buxton: Robert Durban, Arthur Tuffs, and Harry Taylor. Four at Bonny Eagle: Frank Swinington, Sam Berry, Ed Foster and Ernest Foster. Ed Foster and Ernest Foster worked shoving the logs by the Company boom. Four men at Limington Dry Way: Erland Day and Stanley Harnett, Ed Burrill and Kenneth Smith.

June 15
The rear was at Hiland Rips tonight. Four men at Bonny Eagle Pond pushing logs by the Company boom. We did not sluice many as it took all day to fill the pond below the narrows. Something should be done about it before another drive. Men shoving the boom were: E.A. Foster, Wm. Foster, Ed Burrill and Arthur Tuffs. Saw Mr. Hamilton and asked him not to put the boards up at Hiram until we were over the Steep Falls Dam. Four men at Limington Dryway: Erland Day, Stanley Harnett, Kendrick Sawyer and Kenneth Smith.

June 16
Rear still on Hiland. Logs working slow. Twenty-two men on the rear. I found 80,000 of logs in back of the Brick Yard Eddey Island dry that we will have to use horses on. Six men and Foster's team working at the dry logs. Three men at West Buxton sluicing: Sam Berry, Ed Foster and Frank Swinington.

June 17
Trucked Foster's horses to the mill. Three men at West Buxton sluicing: Sam Berry, Frank Swinington and Ed Foster. Rear was at Gould's Island tonight. Three men at West Buxton: Walter Foster, Bob Durban and Arthur Tuffs. Water dropping fast. Was to Union Dam. The C.C. Boom on the Saco side is not out yet. Logs running over Bar Mills good. We had four men at the Limington Dry Way: E. Day, K. Sawyer, Stanley Harnett and Kenneth Smith.

June 18
Three men at Bonny Eagle and West Buxton sluicing: Will and Ernest Foster and Frank Swinington. Six men and team at Eddey at Steep Falls working at the dry logs. Water is dropping. I will keep a crew ahead working of the dry logs. The rear was at the Boy Scout Camp tonight. Saw Mr. Hamilton he says that they will put the boom out at Union tomorrow. Jam at Bar Mills.

June 19
Hauled the boat to Bar Mills. Cleared the jam out and hauled it back to Limington. Worked on a jam at Parker's Rips. Four men and a horse on the dry logs below Brick Yard Eddey. Five men on the rear. It was at Gray's Eddey tonight. Three men at Bonny Eagle shoving by the Company boom. Head wind. Sam Berry, Ed Foster and Frank Swinington.

June 20
Eighteen men at Parker's Rips with the picture men. We lost about a half a day with all the men. Five men working on the rear. We had a jam on Limington. Six men worked all day. Had a jam at Bar Mills. Three men at Bonny Eagle and West Buxton: Ed Foster and Will Foster and Frank Swinington.

June 21
Rear was at the Boom Field. Nine men working on it. Ten men at Limington Dry Way. Logs were held back by the head wind in the dead water above Limington and ran over a body at night. They jammed in the Channel and filled the Dry Way again. Had ten men working at them today. Two men running over Bar Mills: F. Swinington and Ed Foster.

June 22
Rear was over Steep Falls at 2 p.m. We were a day with eight men cleaning the rocks and Eddey under the dam at Steep Falls. These logs ran on the rocks when the Company's boom let go. When their boom was out properly logs did not run on those rocks. Three men at Bonny Eagle and West Buxton sluicing: F. Swinington, A. Tuffs, R. Durban. Eight men working at Limington Dryway.

June 23
Rear was in the Big Eddey below Steep Falls. Eight men at Limington Dry Way. Three men at Bar Mills. Rear moved slowly today.

June 24
Six men working on Limington Day Way. Three men at Bonny Eagle and West Buxton sluicing: F. Swinington, Ernest Foster and Water Foster. Hauled boat back from Limington to the rear. Had about 100 logs in back of brick yard Eddey Island about half afloat. Carried over the bank by the High Water. We had to drag them out. About all the logs in the break away seemed to land back of those two islands. Rear still in the Eddey.

June 25
Six men at Limington rolling dry logs. Had a jam at Bar Mills. Picked it out and ran what logs was in the pond over the dam. Rear was at the rocks below Brick Yard Eddey tonight. All the logs were turned off today stopped at Parker's Rips. Did not sluice at Bonny Eagle on West Buxton today. But did turn over Bar Mills with three men: Erland Day, K. Sawyer and Ernest Foster. Water dropping.

June 26
Rear was at Sanborn Cottage tonight. Six men working on Parker's Rips cleaning dry logs. Three men turning over at Bar Mills. Did not sluice any at Bonny Eagle or West Buxton. Head wind in the pond. No logs came down for two days. Ran a lot over Limington today. Raining but not enough to do any good.

June 27
Ten men working on Parker's Rips. Seven men working on Limington. Head wind all day. Rear around the turn above Parker's Rips. Did not accomplish much on account of the wind. Two men sluicing at West Buxton: F. Swinington and Ed Foster.

June 28
Head wind. Rear was at head of Parker's Rips at 2:30 p.m. Some of crew working on Limington and some on Parker's Rips. Logs moved slow in pond since the boards are up. We've turned a lot over Limington today. Not many came down to the dam.

June 29
Head wind in the ponds. The men working on Limington—balance on Parker's Rips. Was to the Main Boom at Biddeford today. Seemed to be a week's sawing.

June 30
Head wind in the Pond. Just a few logs came down to Bonny Eagle Dam. Rear was over Parker's Rips 3 p.m. Had eight men working on Limington. We sluiced the logs out of the ponds. Three men sluicing at Bonny Eagle and West Buxton: F. Swinington, Bob Durban and Arthur Tuffs. Mr. Hamilton says that he will put the boards on at Union Dam today.

July 1, 1940
Sluicing at Bonny Eagle with 2 men. All the rest of the crew on Limington. Head wind in the pond. Let some of the men go Saturday night. We moved a lot of logs off Limington today. Water fair.

July 2
Fair wind today. Three men at West Buxton sluicing: F. Swinington, Ed Foster and Will Foster. Eighteen men on the Falls. Boulter and the Company clerk seemed to be following the drive checking up on what we're doing. The river below Bar Mills seemed to be picking up the usual amount of logs.

July 3
Three men at Bonny Eagle and West Buxton sluicing. Fair wind in the a.m. Head wind in the p.m. Two men at Bar Mills working on jam below the dam. The Company seems to be holding the water to fill the ponds. Rear below bridge at Limington. Water about the same. Saw Russell Holmes about Limington tax. He says that they will try to collect.

July 4
No men working today. Raining. Was to the dams. They are all full.

July 5
Three men at Bar Mills turning over the dam: Erland Day, K. Sawyer, and Ernest Foster. Two men shoving logs past the Company boom at Bonny Eagle: Frank Swinington and Ed Foster. Three men at West Buxton: Sam Berry, Arthur Tuffs, Robert Durban. Rear in Bonny Eagle Pond 3 p.m. Rained yesterday. Seems to be holding the water up. Leather board not running at Bar Mills.

July 6
Rear out of the Little Ossippi at 3 p.m. Started to clear the big Eddey. Two men at Bar Mills: Day and Sawyer. Logs work bad. The L.B. not running and we don't get the volume of water when they don't run. Two men at the C.C. Boom shoving by at the Bonny Eagle Pond. Four men at West Buxton: F. Swinington, Walter Foster, Ernest Foster and Harry Taylor.

July 7
Had 4 men at C.C. boom in Bonny Eagle till 10 a.m.: E. Day, F. Swinington, Ed Foster, and Stanley Harnett. Three men at West Buxton: Bob Durban, Arthur Tuffs and Walter Foster. Water uniform in the pond today. Rear at the Smith Place.

July 8
Four men at West Buxton sluicing: A. Tuffs, R. Durban, F. Swinington and Ed Foster. Three at Bonny Eagle: Walter Foster, Ed Burrill and Harry Taylor. Two men at Bar Mills turning over: Sawyer and Day. Water fair. The rear at Cook-Ushers Cove tonight.

July 9
Was over Bonny Eagle Dam 2:30 p.m. with all the logs except some boom to take apart. Started at the Stumps in West Buxton Pond. Had 5 men at Bonny Eagle: F. Swinington and E.A. Foster. Two at Bar Mills: Day and Sawyer. Four at West Buxton sluicing: Tuffs, Durban, Ed Foster and Harry Taylor. Leather board. Company will put on boards at Bar Mills tomorrow. Water uniform in the ponds.

July 10
Head wind all day. We cleaned out the stumps and the lower end of New River. Got a boom behind the logs. Had 4 men at Bar Mills: E. Day, K. Sawyer, A. Tuffs and R. Durban. Two rocks below the dam at Bar Mills bother. They should be blown out. We have to maintain two men to keep them clear.

July 11
Rear over the West Buxton Dam at 10 a.m. Hauled the boat and started to clear out. Five men turning over at Bar Mills: F. Swinington, K. Sawyer, E. Day, Stanley Harnett and E.A. Foster. Boards are on at Bar Mills. We have two jams today below the dam. Logs running over Union good.

July 12
Rear was on the gravels below West Buxton. Company gave us water to clean out the back way. Five men at Bar Mills turning over the dam: F. Swinington, E. Day, E.A. Foster, S. Harnett and K. Sawyer. Lot of logs on the gravels but they work good.

July 13
Rear off the gravels at 3 p.m. Five men at Bar Mills: F. Swinington, E. Day, E.A. Foster, S. Harnett, and K. Sawyer. The logs continued to jam on the two rocks below the dam and they would have to be drilled to blow them. Trucked Foster's team to Bar Mills to clean out back of the island. Water fair.

July 14
Rear at Egypt tonight. Head wind all day. Logs move slow in the dead water. Five men at Bar Mills turning over: F. Swinington, E.A. Foster, S. Harnett, E. Day and K. Sawyer. Foster's team and three men on the dry logs below Bar Mills.

July 15
Rear at the head of the piers. Head wind. We had to tow a boom. Five men at Bar Mills: F. Swinington, E.A. Foster, E. Day, S. Harnett and K. Sawyer. Logs work bad today—three or four jams. Water low below the dam.

July 16
Rear into the main body at Bar Mills. Five men turning over at Bar Mills: F. Swinington, E.A. Foster, E. Day, S. Harnett and K. Sawyer. They seem to be holding the water back. The water was running over the Buxton, Bonny Eagle and Hiram Dams but was low below Bar Mills. We had a big jam in Salmon Falls. Took us about one half a day with all the crew. We had a jam at Union by the bridge piers.

July 17
Did not work on the rear. We had three jams in Salmon Falls. Sluiced at Bar Mills in the p.m. Water is good today about a foot more than yesterday. Five men turning over: F. Swinington, E.A. Foster, E. Day, S. Harnett and K. Sawyer. Had a jam at the half way house. Water fairly good.

July 18
Rear was over Bar Mills Dam at 2:30. Hauled the boat and started to clean out under the dam. We had a big jam by the big rock in Salmon Falls. Had to move down with all the crew to clear it out. The logs we turned over yesterday seemed to be all big and not quite water enough to carry them along. They seem to jam everywhere.

July 19
Rear is below the bridge at Bar Mills. Worked one half a day with all the crew on two jams in Salmon Falls. Water condition about the same.

July 20
Rear is at the Island below Bar Mills. Had to wait until 10 a.m. for water to come from West Buxton. We had a jam at Cow Pitch and one at the half way house.

July 21
No one working. The Fiber Company closed the sluiceway today. Sunday no water.

July 22
Rear was at the half way house. We left some logs in back of the island. Will have to get horses to haul them out. Water was low. We did not get enough to drive until 10 a.m. Logs running over Union Dam o.k.

July 23
Rear was on Half Way Rips. We got the water about 10:30. Had Foster's team on the dry logs. We had two jams—one at Cow Pitch and one at the Big Rock. Boulten and the Company Clerk (Haskell) are still with us.

July 24
Rear is at the Big Eddey above the bridge. Foster's team working on dry logs in back of the island. Two jams in Salmon Falls. Finished the dry logs in back of the island.

July 25
Rear was below Salmon Falls Bridge. Had a jam in Cow Pitch. It took one half a day to clean it out. Foster's team and four men working on dry logs in the eddey above Salmon Falls.

July 26
Rear is at the Big Rock in Salmon Falls. The water did not come until 10 a.m. Foster's team and four men at Union on dry logs. Water is low today.

July 27
Rear was at Indian Cellar. Our 42 hours were up at 9 a.m. We worked until 12. Laid off six men: Sawyer, Durban, Tuffs, Swinington, Boulten and Bradbury. Foster's team finished at Union Gravels. Four men worked with him: J.G.D. was up.

July 28
Not working. Sunday. No water.
July 29
Rear was over Salmon Falls at 2 p.m. and we hauled the boats to Pleasant Point. The water was good in Union Pond. We had 15 men working. Head wind.
July 30
Rear was at the narrows in Union Pond. Head wind. We had 15 men working. Water good.
July 31
Rear over Union Dam at 9:30 a.m. and on the gravels at night. Fifteen men working. Water fair. We came good today.
August 1, 1940
We cleared the gravels at 9 a.m. The wind fair. Water good—15 men working. Rear at night where the power line crosses the river.
August 2
We left the rear at noon today above the Fogg Cottage. Head wind— we settled with the crew and pulled the boats on shore.
August 9
Started the rear above the cottages. Rear above Sevey's Bridge at night. Men working: M.E. Cunningham, Stanley Harnett, Kenneth Smith, Carl Bouten, and Frank Swinington. Head wind. Had to tow a boom.
August 10
Rear was below Sevey's Bridge at night. Head wind. Have to tow. Men working were: M.E. Cunningham, Wm. Wintell, Stanley Harnett, Ernest Foster, Harold Burrill, Kenneth Smith, and Erland Day.
August 11
Rear was on Little Falls. Head wind. Men working were: M.E. Cunningham, Stanley Harnett, Ernest Foster, Harold Burrill, Erland Day, Kenneth Smith and Wm. Wintell.
August 12
Rear one half a mile above the water works. We've had head wind ever since we started. Water is fairly good. Men working are: M.E. Cunningham, Stanley Harnett, Ernest Foster, Harold Burrill, Erland Day, Kenneth Smith and Wm. Wintell.
August 13
Rear was at the main body above the Boom and a boom behind the logs. Hauled the boats and stored them. The wind was fair today and the water good. Men working were: M.E. Cunningham, Stanley Harnett, Ernest Foster, Harold Burrill, Kenneth Smith, Erland Day and Wm. Wintell.
Drive Completed
Merle Cunningham, Erland Day, Stanley Harnett and Kenneth Smith taking care of Steep Falls holding boom and picking up the lines.

APPENDIX TWO
A Partial Listing of River Drivers - Truckers - Scalers - Teamsters - Portable Sawmill & Woods Men
J.G. Deering & Son, Saco River Driving Co., Diamond Match Co. and Other Companies
Saco River and Tributaries - 1900 to 1960
("Job Categories," Towns and Dates are Listed to Extent of Available Information)

Name	Town/Relation	Role/Notes
Jere Anderson		River Driver
J.H.A.	Name unknown.	Drive Boss 1905–08 in F.C. Deering's notebook.
Philip G. Andrews	Fryeburg	B. 11/13/1913 Woodsman-Teamster-Trucker
Jay Atkinson	Bar Mills	River Driver
Otis Allard	Otisfield	River Driver
Jack Ballard	Fryeburg	River Driver
Ralph Bean	Steep Falls-Limington Side	River Driver
Everett Bean		River Driver
Walter (Tink) Bemis	Fryeburg Harbor	River Driver
Leo R. Bell	Fryeburg (Brother)	B. 3/3/1889 D. 8/17/1986 Woods & Drive Foreman, Scaler & Timber Buyer, J.G.D. & Son
Ervin Bell	Fryeburg (Brother)	River Driver-Woodsman
Earl Bell	Fryeburg (Brother)	Woodsman
Merton Bell	Fryeburg (Brother)	Teamster-Woodsman
Ralston Bennett	Lovell	B. 12/22/1916 Trucker-Woodsman
Herbert Berry		River Driver
Sam Berry	Bonny Eagle	River Driver
Joe Berry		River Driver
Kenneth Blaney	Cornish	B. 12/24/1910 D. 1/5/1988 Scaler-Clerk-Portable Mill Foreman-River Driver, J.G.D. & Son
Leon Blake	Bridgton	River Driver-Portable Mill Foreman 1955-59, J.G.D. & Son
Everett Bonny		River Driver
Norman Boothby	Saco	Teamster-"Jerry & Ned" and J.G.D. & Son, Biddeford mill crew.
William Howard Boulter	West Buxton (Father)	B. 3/29/1853 River Driver
James Everett Boulter	West Buxton (Son)	B. 4/29/1883 D. 3/2/1944 Woods and Drive Boss Diamond Match Co. following Nelson Nason.
Carl Boulter	West Buxton (Grandson)	River Driver B. 4/8/1874 D. 12/18/1923 Woods and Drive Boss for J.G.D. & Son, 1909-1923.
Frank E. Brown	South Standish (Father)	
Clarence E. Brown	South Standish (Son)	B. 5/29/1904 River Driver-Woodsman
Carl Brown	Lovell	Woodsman
Carl Brown, Jr.	Lovell	Woodsman
Chester Bunker		River Driver
Perley Burnham	Limington	River Driver
Edward E. Burrill	Cornish (Brother)	B. 4/18/1901 - at Limington Bridge River Driver-Woodsman
Roy (Sime) Burrill	Steep Falls- (Brother) Limington Side	River Driver Woodsman
Harold Burrill	Steep Falls- (Brother) Limington Side	River Driver-Woodsman
Leroy Buswell	Fryeburg	River Driver-Woodsman
Wesley Buswell	Fryeburg	River Driver-Woodsman
E. (Ellsworth) Chandler Buzzell	Fryeburg (Father)	Woodsman
Francis G. Buzzell	Fryeburg (Son)	Woodsman
Donald Buzzell	Fryeburg (Son)	Woodsman
Walter Casey	(Born Eagle Lake) E. Baldwin and Cornish	River Driver
Wilfred H. Charron	Sebago Lake	B. 11/2/1916 Woodsman-Teamster
Edmund Maxwm Chisholm	Kezar Falls (Brother)	B. 1/10/1915 D. 3/2/1988 River Driver-Woodsman
Russell Chisholm	Cornish (Brother)	B. 1916 Drowned 5/23/1940 on the drive below mouth of Pleasant Pond
Mark R. Coolbroth	Steep Falls-Standish Side	B. 9/18/1827 Timber Buyer and Drive Foreman for J.G.D. & Son, Hobson and others.

It is our intent to add to this list as an historical record. Corrections, additional information on men listed and new names are welcomed, as are photographs and memorabilia of Saco River lumber history. Please write to T.M. Armstrong, 14 Elm Street, Biddeford, ME 04005.

Caleb Cousins	Standish (Father)	River Driver
Lawrence K. Cousins	Limington (Son)	B. 3/25/1904
		D. 7/31/1961
		Drive Foreman 1943, J.G.D. & Son,
		River Driver-Woodsman and Teamster.
Asa C. Cunningham	Saco/Gorham (Father)	B. 12/27/1885
		Stillwater, ME
		D. 1/17/1968
		Woods and Drive Boss for
		J.G.D. & Son, 1926-1943.
Merle E. Cunningham	Saco (Son)	B. 12/12/1910
		D. 6/24/1973
		River Driver-Woodsman-Trucker
		Drive foreman late 1930's
Charles Day		Lovell Portable Mill
		1943
		Teamster-Woodsman
Calvin Day	Bar Mills	River Driver
Erland E. (Babe) Day	Cornish	B. 1911
		D. 3/23/1978
		Woods and Portable Mill Foreman-
		J.G.D. & Son
		River Driver-Woodsman
		Drive Foreman 1941?
Newt Day	Hiram	River Driver
Philip Day	Limington	Woodsman
Raymond Day	Cornish	Woodsman
Everett Davis		River Driver
Howard Davis	West Buxton	River Driver
Dudley Davis		Lovell Portable
		Mill 1943
Robert Dearborn	E. Limington	B. 10/11/1904
		D. 9/23/1962
		River Driver-Woodsman
George Dyer		River Driver
Jack Eastman	Lovell	Woodsman
Lawrence Eastman		Woodsman
Louis Emmons		Teamster
Richard Fleck (?)		River Foreman-
		Androscoggin Pulp Co.,
		Steep Falls 1920's.
Alonzo (Lon) Foster	Steep Falls	B. 1/12/1887 West Buxton
		D. 10/27/1964
		River Driver-Drive Boss,
		J.G.D. & Son 1925
Edwin A. Foster	Steep Falls-	River Driver-
	Limington Side	Woodsman-Trucker
Ernest Foster	Steep Falls-Denmark	River Driver-Teamster
Walter Foster		River Driver
Will Foster		River Driver
John Fox	Lovell	River Driver-Woodsman
		Sawmill Operator
Kenneth Frazier	Fryeburg	River Driver
Sidney Frost	South Hiram	Woodsman
Ernest Gagnon	Biddeford ?	River Driver
John (Frenchy) Gallant	Steep Falls ?	River Driver
		B. 4/12/1912
Harold C. Gain	Fryeburg	Woodsman-Trucker
Frank (King) Graham	Fryeburg ?	River Driver
Albion Gordon	Fryeburg	Timber Buyer-
		Land Surveyor early
		1900's, J.G.D. & Son
Lowell Graham	West Buxton	River Driver
Albert Graves		Trucker
Lawrence ? Gray	Fryeburg	Diamond Match Co.
		Foreman
Simon Guptil	Denmark	Scaler
Bob Haley	Limington	River Driver
Chester Haley		Woodsman
Harry Haley	Limington	River Driver
Ralph Haley	Limington	River Driver
George Haley	Limington	River Driver
Owen Hall		River Driver
Stanley (Shorty) Harnett	Cornish	D. about 1973,
		River Driver-
		Teamster-Woodsman
Earl Harriman	Fryeburg Harbor	Lovell Portable
		Mill 1943
Roy Higgins	West Buxton	River Driver
William (Father) Hodgeton	Steep Falls	River Driver
Freeman Howard	Hiram	River Driver-Woodsman
Harry Howe	Steep Falls	River Driver
Alfred Hubert	Biddeford	B. 12/27/1895
		D. 1967
		River Driver-J.G.D. & Son
		Biddeford Boom-Mill Crew
Win Hutchins		Teamster
Lionel Hutchins	Hiram	River Driver
Earl Johnson	Hiram	River Driver
Ike Johnson		Woodsman
Skiff Kelso	Bar Mills (Father)	River Driver -
		Diamond Match Co.,
		Foreman 1920's
Elmer Kelso	Bar Mills (Son)	River Driver
Perley Keniston		Woodsman
Ford Keniston	Lovell ?	Woodsman
Jason Knights	Hollis ?	Woodsman
Herbert Kimball	Fryeburg	River Driver
Virgil Kiesman	Fryeburg	Woodsman
Joe Lacroix	Biddeford	Teamster-"Joe &
		Charley" and J.G.D. & Son
		Biddeford mill crew
Chester Leonard	Fryeburg	Lovell Portable
		Mill Foreman, J.G.D.
		& Son, 1952-53
Robert Littefield	Lovell	B. 6/7/1903
		River Driver-Woodsman
Walter Lord	Fryeburg	Woodsman
S.R.L.	Name unknown	Drive boss 1903-04 in
		F.G. Deering's Notebook
Harry Lord		River Driver
Scott Macorison	West Buxton	River Driver
Gerard Martel	Alfred	River Driver
Abraham (Bram) Martel	Biddeford	River Driver
Israel Martin	Cornish	Woodsman
John Martin	Cornish	Woodsman
Roy Martin	Cornish	Woodsman
Billy Mason		River Driver
Joe Mason	Bar Mills	River Driver
Herbert McKinney	Gorham	Woodsman-Teamster
Richard Merrifield	Steep Falls-	River Driver-Woodsman
	Limington Side	Drive Foreman 1942 J.G.D. & Son
Alphonse Morin	Biddeford	B. 8/18/1898 Decd.
		Biddeford Boom-Mill Crew
Hilaire (Eli) Morin	Saco (Father)	B. 8/6/1884
		D. 2/14/1962
		Woodsman- J.G.D. & Son
		Mill Crew
Ralph Morin	Saco/Sanbornville, N.H. (Son)	B. 3/16/1918
		Woodsman-J.G.D. & Son
		Mill Crew & Sales
Carleton Nason	Steep Falls	River Driver
Everett Nason		River Driver
Joe Nason		River Driver
Nelson Nason	E. Limington	B. 1853 D. 1939,
		Drive Boss for
		Diamond Match Co.
		early 1900's
Walter Nutter	Fryeburg	Scaler
Charles Parker		Teamster-Woodsman
Lewis (?) Pendexter		Timber Buyer, J.G.D. & Son 1920's
Carroll Perkins	Cornish	B. 4/9/1913
		River Driver-Teamster-
		Trucker-Woodsman
René (?) Plourde	Biddeford ?	River Driver
Martin Qually (?)	Lovell	Woodsman

Name	Location	Role
Theodore Randolph		River Driver
Harlan (Hal) Richardson	(Born Island Falls) Steep Falls	D. 1979 River Driver
Guy Ridlon		River Driver
Herbert Ridlon	West Buxton	River Driver
Theodore Ridlon		River Driver
Ambrose Rose		Woodsman
Charles Sawyer		River Driver
Kendrick Sawyer		River Driver-Woodsman
Millard (?) Sawyer		River Driver
William Smart	Lovell	River Driver
Ed Smith	Denmark	Woodsman
Ervin (?) Smith	Lovell (Father)	Woodsman-Teamster
Leroy Smith	Lovell (Son)	B. 12/4/1908 River Driver-Woodsman
Kenneth (Buster) Smith	North Fryeburg (Son)	River Driver-Woodsman
Norman Smith	Fryeburg	Teamster-Trucker
Herman H. Stiles	Biddeford/Saco ?	B. 1/7/1898-Decd. Scaler-Grader-Woodsman
Frank Sturgis		Woodsman
Herbert Sturgis	Standish	River Driver
Wendell Sturgis	Standish	River Driver-Woodsman
Earl Southwick	Bonny Eagle (Father)	River Driver
Thomas Southwick	Bonny Eagle (Son)	River Driver
William Southwick	Bonny Eagle (Son)	River Driver
Frank Swinington	Steep Falls	River Driver-Woodsman
Harry (Sam) Taylor	Steep Falls	B. 2/17/1906 D. 6/26/1992 River Driver-Woodsman
Charles Tenney	Denmark	River Driver
? (Jumper) Thompson	Bar Mills	River Driver
Herbert Towers		River Driver
David Tripp		River Driver
Arthur Tufts		River Driver
Cook (?) Usher		
Ellis Usher }	Born Hollis, lived in Limington (twin brothers)	B. 2/2/1887 Decd. River Driver-Woodsman
Gilbert Usher }		B. 2/2/1887 D. 7/5/1975 River Driver-Woodsman
Fred Usher		River Driver
Joe Wallace	Bar Mills	River Driver
Frank (Long) Walker	Fryeburg	River Driver
Kenneth Walker	Saco	B. 2/24/1914 Portable Mill Foreman, J.G.D. & Son, 1952–55
Eddie Watson	Waterboro	River Driver
Robert Watson		Lovell Portable Mill 1943
Verd (?) Waterman		Woodsman
Neil C. Warren	Steep Falls	B. 10/6/1895 D. 9/12/1976 River Driver-Woodsman
Ivan Warden	Steep Falls-Limington Side	Woodsman
Percy Weeman	West Buxton	Teamster
Hal Whitehouse	West Buxton	River Driver Drive Boss J.G.D. & Son, 1924
William Wintell	Cornish	River Driver
James Wood		River Driver
? Woodbury	Sebago	River Driver
Charles Yates		Teamster

APPENDIX THREE
My Dad Was A River Driver—By Clarence E. Brown
(About his father, Frank E. Brown—Woods and Drive Boss for J.G. Deering & Son who died in 1923 at age 49.)

Dad left home early in the spring as soon as the ice was out up river. We carried him up to Steep Falls, where he boarded the train and went to Fryeburg. Harry Holt met him at the depot and they went to an old house where they boarded themselves. There were about twenty-two to twenty-six men in all; one crew worked for J.G. Deering & Son, and the other worked for the Diamond Match Company of Biddeford. A cook and helper were included in this group and all were equipped with rubber boots and rain coats.

We worked every day rain or shine, walking to work through the woods and the meadows. Sometimes when the water was high, we used the batteau, a big boat with flaring sides and pointed at both ends. We used four oars, two on each side and a man in each end to steer it.

After the logs were all rolled in the current carried them down stream. High water over-flowed the banks and some of the logs would go over the banks into the woods. It was our job to get them back into the river again. Four to fourteen men would grab onto a log with cant-dogs and catch it to the water, wading in the water with muck ankle-deep or more in some places.

Next we went to Lovell where the logs went over a dam into Kezar River, drifting about a mile or more to where the water was over-flowing into a big meadow. We put them into a boom made of logs "toggled" together. We then made a raft by toggling six or eight big logs together. In the middle of the raft a wooden post was set over which stood on end a winch was attached a rope 1 inch in diameter and a hundred and fifty feet long, with an anchor on the other end. We carried this anchor as far as we could out in the meadow and dropped it, then went back to the raft and started winding the rope on the "capstan", as it was called; pulling the raft of logs across the meadow until we came to the anchor. We then pulled the anchor into the boat. This maneuvre was repeated until the boom of logs had crossed the meadow and into the old river, into the current where they would go down by themselves. We followed along, rolling in the logs that stopped on sand bars, stumps, fallen trees and whatever was in the way. There were rapids with rocks of all sizes to stop the logs and make "log- jams." Where the rapids were not too steep we followed in the boat. As we approached a "jam" we got out of the boat, moving to the down-river end, moving one log at a time until the "jam" started to move. Sometimes when we moved the right log the whole "jam" would flatten out and there would be a mad scrabble for the boat. This was repeated until all the logs were off the rocks. Going down river we passed through Goose Gall Eddy, Walker's Island, Hiram Falls, The Wife, Sokosis, Steep Falls, Parker's Rips, Limington Falls and Limington Eddy where the Little Ossipee River flows into the Saco. From Killick Brook to Bonny Eagle Dam we had to make a boom across the river on account of the dead water; no current to speak of and the wind would blow the logs back up river.

At Bonny Eagle Power House we had to take turns cranking up the gate at the head of the sluice, standing on a boom log and steering the logs into the sluice with long pike poles all day. Before we left at night we cranked the gate shut to save water. This procedure was followed until all the logs were through the gate which was then closed.

The next dam was at West Buxton where another log sluice with a gate to open and close was located. Below the bridge at West Buxton was a large sandbar. This would cover with logs. Percy Weeman was placed here with a pair of horses and a mule to help get the logs back into the water. Roff's Island came next with more sand and a long stretch of good driving. We then came to Bar Mills Dam where the logs went over the middle of the dam, a section of flush boards having been removed for this purpose. More rocks were passed as we started down through Salmon Falls, passing through some of the worst rough water on the river; a deep gorge in solid ledge, steep sides and swift current. Just below the bridge was a big rock in the center of the gorge with water around it. The logs jammed and filled the river at this point. After most of the logs had been removed they left two men to get the last of the jam off. These men had ropes tied to them so as to get them ashore. These ropes helped them to swim ashore.

Lawrence Cousins fell off a jam at Cow Pitch and went down the rest of the way to the eddy below the falls. He was the only man to survive this test. He told me that at first he was rolling and tumbling with the water, then managed to come to the surface and get a breath, rolling and tumbling again until he surfaced in the eddy. His head came up between two logs and he put an arm over each log which saved his life.

We are now below Salmon Falls and passing through a stretch of slow water. We come to another dam at Union Falls, the last falls on the river until we come to Saco, where the logs are sorted. There is a gap in the boom where every single log has to be checked for the company's mark. J.G. Deering's went on the Saco side of a boom down the center of the river. The Diamond Match Company went on the other side of the river. Here the logs were hauled out of the water into the mill.

The mill was equipped with band-saws. A man rode on the carriage and "dogged" the logs so they couldn't turn while being sawed. It was hard and dangerous work. One man rode a half hour then another took over, the work being too hard for one man to work steadily all day.

The lumber from Deering's logs was sawed into boards and dimension timber to build with. The logs from Diamond Match Company were sawn into plank and dried then cut up into match blocks. Being clear between the knots, these blocks went to another mill to be made into matches. These matches were shipped all over the world.

Just to get back to another important part of the River Drive, we will comment on the blacksmiths. There was one in each of the following towns: Fryeburg, Hiram, Baldwin, Steep Falls, East Limington, West Buxton, Bar Mills and Saco. Before starting a drive all broken peavey handles had to be replaced, all "dogs" were reshaped, sharpened and tempered and picks in the cantdogs drawn out and squared so they would hold while pulling a log along into the water.

The long pike poles [*pickpoles*] came without ferrules or picks which had to be installed, sharpened and tempered. There were hundreds of toggle-chains (a piece of chain about two feet long with a piece of round iron rod bent to form a "U"), put through each end link, welded together and drawn to a point to be driven into a log; the other point into another log, making a boom, or chain of logs which was used in many ways. This was stretched across the river behind the logs to keep them moving in slow water. They went ahead of the main drive and made booms across the large coves and small streams to keep the logs out. They steered them into the sluices in the dams. In eddys the current would keep the logs going around and around. We used a boat on one end of the boom-row up river and behind the logs, then down until they would be themselves.

There was one big eddy just below Walker's Bridge, followed by a little set of rapids. We would get the rear over the rapids about noon-time. There being thousands of logs in the eddy, Dad would tell the drivers they may as well go home. The next morning when we came to work about all the logs were gone from the eddy; we had a half day off and the whirling water had done the work.

Let us not forget the boarding houses. The old Fred Holt House where the men boarded themselves, Walker's in Fryeburg, the Liberty House or Hotel in Brownfield (where Dad upset the blueberry sauce in the middle of the table and blamed it on Father Hodgdon), West Baldwin at Frank Hodgdon's (one of the best cooks on the river), Randall Foss, East Limington (just an old farm house with some of the beds upstairs in an open attic. It was here that Bert Towers threw the calk shoe at Father Hodgdon, just missing his head but the calks made scratches on his nose) and Ethel Potter's at West Buxton (a widow with four girls and a boy). After supper the gang played horse-shoes or played the phonograph. King Graham had an accordion which he played. He also went to Wilbur Townsend's barber shop and had his head shaved, getting a sunburn which made him a red-head! Anderson's, in Dayton, another farmhouse was the last place we boarded on the way.

There is a statue or cut-out of a River Driver standing on a log in the park near the new shopping center in Saco. This statue has calk shoes, felt hat, and leaning on a cantdog stands about twelve feet tall. It was made by Joe Deering and is taken from a snapshot of my Dad (*see photo on page 14*).

[According to Clarence Brown's family, Clarence wrote this piece about 1970.]

APPENDIX FOUR
A Short Story on Saco River White Pine
by Frank C. Deering
Spring, 1901

We are most anxious to say a word to you about the husbanding of your pine lands and the need of some application of the present forestry methods in vogue, to increase and perpetuate the stumpage of pine, hemlock and spruce in the Saco River valley. As you well know, lumbering has been carried on up and down the length of Saco River for more than two hundred and fifty years, and during that long stretch of time the manufacturers of pine, spruce and hemlock have paid in to you many millions of dollars. Take the valleys of the Little Ossipee, Great Ossipee and the many smaller streams flowing into the Saco, and the volume of timber that has been driven forth from these streams every spring during the time mentioned becomes a return to you which, invested when you have few other duties, and a sure and quick return in dollars and cents after it is harvested. Again, having seeded one small corner of your land, hunt for another, seed that, and keep your crops seeding so in the time mentioned you will have a crop of pine for each few years, and just as sure and regular a return as your hay, grain and vegetables.

Assuming that you have an old pasture or field that nature has seeded for you, don't leave it as I have seen so many nurseries, so crowded with trees that the land is sapped and the growth is slow. Take a part of the fall and winter and thin it out and let in little sun, and just as sun and food make healthy boys and girls, so will sun and nature's food make healthy, swift-growing trees.

As to trimming, authorities differ; but I believe a judicious removal of limbs drives the growth into the trunk and in no way impairs the health of the tree. Too heavy tops shut out sun, and God's warm sun is just as essential to the good growth of a tree as plenty of it shining through the windows of your house is to the happiness and health of your household.

J. Sterling Morton, Ex-Secretary of Agriculture, writing on "The Vital Importance of Forests" in *The Saturday Evening Post*, says:

"The whole future of agriculture is vitally dependent upon arboriculture and forestry. The reckless destruction of the groves and forests of the United States threatens utter infertility to all parts of the Union. More than twenty-five thousand acres of trees are cut down in the United States and made into railroad ties, lumber, furniture and other commodities every twenty-four hours! Among seventy-five millions of people even this vast destruction of trees gives a very small per capita portion of wood or lumber. The interdependence of tree life, vegetable life and animal life is constant. Unless forests are conserved and trees planted, all farming must perish within the next hundred years; and should the whole globe be denuded of forests and groves all animal life would become extinct. The intermission of tree life and growth throughout the world, for a single summer, would extinguish all animal life. Teaching tree-planting at home, about the fireside, and practicing tree-planting on all the farms in the United States, are pressing necessities. Self-preservation should inspire every American to do all in his power to promote arboriculture. Agriculture everywhere must lend a helping hand to the tree-planters. It is time to plant trees and begin the partial renewal of the forests of the continent, if we care to leave to our posterity a habitable country."

This may sound like a warning that pine is fast disappearing on the Saco River valleys. Such is not the import of this article however. Just as many acres are covered with growth today as fifty years ago. The character has change, however, and it is to prevent further changes and to perpetuate our reserve that we ask you to carefully consider what we write.

We are a progressive nation, quick to see in which direction our advantage lies, and just as quick to follow it. We see everything being done with system. Think of the marvelous system required to control and operate the recently organized Iron and Steel Company, with a capital of over a billion dollars, and remember that the same fine system, the same careful weighing of the right way to operate and the wrong, the same quick energy of today, apply just as advantageously to your smaller operations as to these great industries.

There is no man earning wages or managing corporations so independent as the New England farmer. Only apply your leisure moments to the study of when, where and how to market your produce and increase the burden of your crops, and you enjoy greater freedom and better returns for labor expended than most of your associates in the struggle for life. It is always the man who applies careful system to the production of his farm or his factory, who patiently studies new methods for increasing the output of his farm or his factory, that achieves success.

In the interest of good fellowship between the man who owns the stumpage and the man who buys and manufactures the logs, we have written this short article, and we sincerely hope it will receive more than a passing notice from you. Keep your pine growing. Keep it ahead of you all the time. Look carefully after the growth you already have, and the agitation now going on in other states and other parts of this state with regard to our fast- disappearing forests need not apply to the Saco Valley.

You can take a just pride in having furnished timber for the earliest mills in America, and if you follow the system and care of the present you can take equal pride in handling to your succession the foundation for being the last.

Don't think that a crop which will not grow and mature in one short season is not worth planting. Neither argue that a crop planted in the evening of life cannot be harvested by you and hence matters not. The greatness and richness of this country is wholly due to men who lived and labored for the future and who seeded what we are reaping, fully aware that the fruits of their labor must fall to other than themselves, yet striving on with the future of this great country constantly before their eyes.

Two hundred and fifty years of an honorable industry is a record that should appeal to every farmer and mill owner on Saco River and its tributaries, and it is with the hope that the dawn of this glorious century may also be the dawn of another two hundred and fifty years for this lumber-producing valley, that we address to you this little pamphlet.

APPENDIX FIVE
A Short History of J.G. Deering & Son
by Joseph G. Deering

Our own business was started in 1866 by my grandfather, for whom I was named, through a set of circumstances which he certainly did not anticipate.

He had come from Waterboro and, as a boy worked in a sawmill on Cook's Brook, located on a piece of land which currently belongs to the Federal Forest Service. One of a large family, he had left home and, as a comparatively young man, had gone to sea. He hadn't liked it and came back to Saco, and somehow got into the grocery business. He ran a grocery store at the present location of the Harper Grain Company (Pepperell Square, Saco). He seemed to have a good deal of natural trading ability. Apparently his store was successful, and he gradually accumulated some surplus money.

The pine timber in the Saco Valley had always been an alluring thing and somebody, whose name I don't know [*Living Lane*], persuaded my grandfather to loan him some money [*in 1866*] to build a sawmill beside a dam [*Bradbury*] connecting Spring's Island with Gooch's Island. As sometimes happens, the original amount loaned was not enough, and Grandfather had to put in more and more until it finally proved that he had to take it over and start out in the lumber business. So, you can see, quite unintentionally, he got into the lumber business and—probably quite intentionally—my father, Frank, was born the same year.

In the early days the business was confined to making heading for sugar barrels and sugar boxes—both of which, in my understanding, were shipped green to the West Indies, and must have been in rather sad condition by the time they were unloaded in that hot climate.

My father's education was rather short-lived. His first year in the old Saco High School, which is now the Jordan School next to the railroad on School Street and bordering on Pepperell Park. He and another boy—arriving at school early one morning—had the imagination and lack of sense to hoist a hound dog up through a scuttle which was over the schoolroom and await events. When the hound finally woke up he made considerable disturbance in his elevated position. School had to be dismissed, and Father and the other boy were permanently relieved of having to go to school any more. As sometimes happens, there was a rather frank discussion between father and son and an agreement reached that school should be given up and my father would go to work. I tell you this because I think my father, through his own efforts afterward, became a very highly educated man through reading and study.

The grocery business, of course, is the one which produced money. On Main Street there were two grocery stores: one run by Pierpont Jordan in Biddeford, and one run by my grandfather in Saco. Grandfather, seeking to install his son in the most profitable business, built the store at the corner of Main and Jefferson Streets, which was for so many years occupied by Andrews & Horrigan, now used largely by the Potter Furniture Company.

When I first came back here to work [*1919*] one of my immediate jobs was the collection of rents: from Carlos McKenney on the corner; Jimmy Wood's plumbing shop next door; and then, the double store of Andrews & Horrigan. There were some rented rooms on the second floor and the top floor was occupied by Jimmy Carroll and a poolroom. It was a miserable job. The tenants either didn't have the money, or the ceiling leaked, or there was something wrong with the stairs. I presume a large number of my white hairs came from early efforts of trying to ponder frictions which came in that block.

Apparently, again there was a matter-of-fact talk between father and son. My father announced that he didn't want to go into the grocery business, and he didn't.

There were a couple of small water mills on either side of the dam along Elm Street, between Spring's Island and Saco, and he and Grandfather— sometimes together and sometimes separately— operated the three mills. From the beginning in 1866 they, at first, were unable and then unwilling to extract themselves from this fascinating business. Logs were cut farther and farther up the river and, with everyone continually prophesying there wouldn't be any timber in a few years, this process has continued for over three hundred years. I don't believe there is a river in the world that has a lumber history of as long duration as the Saco River.

The business changed a good deal in my father's time [*incorporating 1903*]. A competitor, the Hobsons, was finally bought out by the Portland Star Match Company, which was afterwards, in turn, bought out by the Diamond Match Company; but the retail business which had been part of the Hobson operation was sold to my father.

The shipping of sugar boxes and headings, which were shipped by schooner to the West Indies, disappeared. Gradually a whole new series of customers and methods came into being.

When I came back here we ran two log drives—an early one in the Spring, and a later one in mid-summer. Occasionally, some logs would be left hung up and frozen in through the following winter. The river drivers were paid $1.00 and $1.25 a day and boarded. They went to all the ball games up and down the river and, when they worked, they worked from daylight to dark. In the early spring they were sometimes in ice water up to their waists.

From this beginning the business has developed to its present situation. It now manufactures finish lumber, moldings out of Maine grown Eastern White Pine and Hemlock which it retails and wholesales locally and to several other New England markets. In addition to its own products, it sells materials such as roofing, shingles, doors, windows, plywood, hardware, insulation and other items used in the construction field. These products come from all over the United States and Canada.

Joseph G. Deering
November 8, 1956

NOTES

INTRODUCTION

1. Roger E. Mitchell, "Occupational Folklore: The Outdoor Industries," in Richard M. Dorson (ed.), *Handbook of American Folklore* (Bloomington: Indiana University Press, 1983), p. 133. The book mentioned is Roger E. Mitchell, *"I'm A Man That Works": The Biography of Don Mitchell of Merrill, Maine* (Orono: *Northeast Folklore* XIX: 1978).

2. Roy P. Fairfield, *Sand, Spindles and Steeples: A History of Saco, Maine* (Portland: House of Falmouth, 1956), p. 107.

3. Deering, written memo, circa 1947, located in Deering Papers, Dyer Library-York Institute Museum, Saco, Maine.

4. Fairfield, *Sand, Spindles, and Steeples*, p. 107.

5. Fairfield, *Sand, Spindles, and Steeples*, p. 106.

6. Susan D. Jones, *Deering Family of Southern Maine* (Portland, 1979), p. 118.

7. Jones, *Deering Family of Southern Maine*, p. 120.

CHAPTER ONE "AN OVERVIEW OF THE LUMBER INDUSTRY OF THE SACO RIVER"

The author would like to acknowledge the research efforts of Richard Roney and Marjorie Hartman which helped make this chapter possible.

1. Charles T. Libby, ed., *Province and Court Records of Maine* (Portland: Maine Historical Society, 1928), Vol. I, pp. 96-102.

2. Sybil Noyes, Charles T. Libby and Walter G. Davis, eds., *Genealogical Dictionary of Maine and New Hampshire* (1928-1939; reprint, Baltimore: Genealogical Publishing Co., 1979), pp. 184-5. Paul J. Lindholt, ed., *John Josselyn, Colonial Traveler: A Critical Edition of Two Voyages to New England* (Hanover, University Press of New England, 1988), p. 140. William Hubbard, "A Narrative of the Troubles with the Indians in New England from Piscataqua to Pemaquid," in *The Indian Wars of New England*, ed., Samuel Drake (New York: Burt Franklin, 1971 orig. publ. 1677), pp. 105-10.

3. George Folsom, *History of Saco and Biddeford* (Saco: Alex C. Putnam, 1830), p. 208; Sybil Noyes, Charles Libby and Walter Davis, eds., *Genealogical Dictionary of Maine and New Hampshire*, p. 699.

4. Folsom, *History of Saco and Biddeford*, pp. 207-8.

5. Folsom, *History of Saco and Biddeford*, p. 208.

6. Folsom, *History of Saco and Biddeford*, p. 256; Neil Rolde, *Sir William Pepperrell of Colonial New England* (Brunswick, Maine: Harpswell Press, 1982), pp. 27-30.

7. Folsom, *History of Saco and Biddeford*, pp. 243-6.

8. James Sullivan, *History of the District of Maine* (Boston: Thomas and Andrews, 1795, p. 25. See also Martin H. Jewett and Olive W. Hannford, *A History of Hollis, Maine 1660-1976* (Farmington: The Knowlton and McLeary Co., 1976), pp. 21-22, 38; Robert C. Taylor, *A History of Limington, Maine 1668-1900* (Oxford: Oxford Hills Press, 1975), pp. 88-93; George Varney, *A Gazetteer of the State of Maine* (Boston: B.B. Russell, 1881), pp. 429, 456.

9. Folsom, *History of Saco and Biddeford*, pp. 260-2.

10. Folsom, *History of Saco and Biddeford*, p. 308.

11. Folsom, *History of Saco and Biddeford*, p. 309.

12. *History of York County, Maine* (Philadelphia: Everts and Peck, 1880), p. 366.

13. *Maine Democrat*, March 25, 1851, p. 3.

14. Roy P. Fairfield, *Sand, Spindles and Steeples: A History of Saco, Maine* (Portland: House of Falmouth, 1956), pp. 104-05.

15. Fairfield, *Sand, Spindles and Steeples*, pp. 105-6.

16. Fairfield, *Sand, Spindles and Steeples*, pp. 106-7.

17. David C. Smith, *A History of Lumbering in Maine 1820-1861* (Orono: University of Maine Press, 1972,) pp. 41, 69.

18. Fairfield, *Sand, Spindles and Steeples*, p. 107.

19. Fairfield, *Sand, Spindles and Steeples*, pp. 108-9.

CHAPTER TWO "TIMBER CRUISING, OPERATING TIMBER LOTS, AND SCALING"

1. Richard G. Wood, *A History of Lumbering in Maine, 1820-1861* (Orono: The University of Maine Press, 1935, reprinted 1972), p. 13.

2. Robert Pike, *Tall Trees, Tough Men* (New York: W.W. Norton & Co., Inc., 1967), p. 197.

3. Hayden L.V. Anderson, *Canals and Inland Waterways of Maine* (Portland: Maine Historical Society Research Series, 1982), pp. 47-57.

4. Deering, written memo, circa 1947, located in Deering Papers, Dyer Library—York Institute, Saco, Maine.

5. Deering, correspondence to Herbert Locke, September 1943, located in Deering Papers, Dyer Library—York Institute, Saco, Maine.

6. Deering, correspondence to Fred Gordon and William Wyman, May 1947 to February 1948, located in Deering Papers, Dyer Library—York Institute, Saco, Maine.

7. For more on the fires of 1948, see Joyce Butler, *Wildfire Loose*.

8. Report of the James W. Sewall Company entitled "Report on White Pine Tributary and Adjacent to the Saco River Valley in Maine and New Hampshire, 1943, for Mr. Joseph G. Deering." Located in Deering Papers, Dyer Library—York Institute, Saco, Maine.

CHAPTER THREE "THE SACO RIVER DRIVE"

1. David C. Smith, *A History of Lumbering in Maine, 1861-1960* (Orono: University of Maine Press, 1972), p. 64.

2. Edward D. Ives, *Argyle Boom* (Orono, *Northeast Folklore* XVII: 1976), p. 18.

3. Edward D. Ives, *Argyle Boom*, p. 26.

4. Edward D. Ives, *Argyle Boom*, p. 30-31.

5. Edward D. Ives, *Argyle Boom*, p. 20.

6. Deering Papers, Dyer Library—York Institute, Saco, Maine. The 1805 ledger and journal of the "Proprietors of Saco Boom" contains these names.

7. David C. Smith, *A History of Lumbering in Maine*, p. 69, and Portland Eastern Argus, June 20, 1879.

8. David Smith, *A History of Lumbering in Maine, 1861-1960*, p. 41-69.

9. Walter Wells, *Water Power of Maine* (1869), pp. 72-74.

10. Walter Wells, *Water Power of Maine*, p. 75.

11. Edward D. Ives, *Argyle Boom*, p. 15.

12. *Fryeburg Historical Newsletter*, Volume I, No. 2 (April 1992), pp. 1-3.

GLOSSARY

Batteau: A double-ended sharply-pointed rowing boat of shallow draft, usually thirty-two feet in length with a six to seven foot beam, used for log driving and freight transport.

"Black Growth": Coniferous trees—pine, spruce, hemlock, fir, etc.

Boom: A series of long logs fastened end to end to guide the direction of or contain logs and pulpwood in the water. Logs were also so enclosed in an area known as the boom.

Cant Dog: A tool with strong wooden handle three to six feet long— similar to a peavey but having a blunt tip with a "toe," and an iron ring around it, instead of a spike. It was used for turning, rolling logs.

Carriage: The moving frame of the sawmill which carries the log firmly in place as it moves against the head saw.

Calked Boots (or caulked): Specially designed boots with steel calks screwed to the sole. Used for walking on floating logs or slippery surfaces.

Cribwork Pier: A structure made of logs fitted at horizontal right angles. Resting on the bottom, rising well above water surface—filled with rocks—piers may also have additional protective vertical logs or sheathing applied. Used to anchor booms, sluice gates and dams or to direct water or log flows.

Cruiser or Timber Cruiser: The person who estimates and/or purchases stumpage or timberland.

Dogger: In a sawmill, the person clamping the log firmly into place on the carriage.

Driving or "On Drive": The transportation of logs by floating across a lake or downstream. The term may refer to the operation or to the wood itself.

Dryway: The part of the waterway that may dry up when water levels drop.

Edger: In a sawmill the machine (or person running the saw) which cuts the lengthwise edge of a board removing bark and/or defect.

Freshet: A high or sudden flow of water above normal levels.

Headsaw: The principal saw in a sawmill.

Landing (or yard): A location, in or near the timber, where wood products are gathered for further transport (formerly water, now by vehicle).

Lignisan: A chemical now or formerly used to prevent blue stain of unseasoned lumber.

Logan Hole: A cove in a river or a mouth of a smaller stream (where logs on the drive collect).

MBF: Abbreviation for thousand board feet—a standard unit of volume measurement for logging and milling. One board foot is one inch thick x 12 inches wide x 12 inches long.

Master Driver: River drive superintendent or boss.

Molder: A wood working machine to shape wood into decorative moldings and trim of all sorts.

Operator (or Contractor-Jobber): A person or company responsible for cutting and hauling a specific timber operation.

Peavey: Now a generic name (often used interchangeably with cant dog), it is named for the inventor—in 1857—Joseph Peavey of Bangor, Maine. A tool with strong wooden handle three to six feet long, with a sharp steel spike at one end and an adjustable sharp steel hook. Used for turning, rolling logs.

Pickpole (or Pike Pole): A long wooden pole (ten to sixteen feet) tipped with a sharp steel point. Used to guide floating logs.

Planer: A woodworking machine to shape and finish various sizes of "rough" wood to uniform size. For example, timbers, tongue & groove boards, and square edged boards.

Re-saw: A woodworking machine of single or multiple saws that reduces larger timbers or cants to smaller size.

Rear: The trailing end of the log drive. While the main body of a drive moves downstream, a crew stays with the rear to move along stranded logs.

Roll in (or turn in/over): The process of putting logs into the water from the banking.

Sawyer: The person controlling the head saw at the mill.

Scaler (or surveyor): The person measuring forest products accurately for length, volume and value.

Setter: In the sawmill, the person controlling the carriage so that the log passes through (lengthwise) the head saw to produce the required thickness of lumber.

Skidway: A raised structure of logs or earth from which logs may be rolled on to a vehicle.

Sluiceway: A passage allowing logs to pass through or over a dam or other obstruction in the water.

Sorting Gap: The location at the boom from which logs are identified and segregated.

Stumpage: Standing wood or timber (not the land itself) measured in volume units or dollars.

Trimmer: In the sawmill, planing mill or lumber yard, the machine or person cutting lumber to certain lengths.

Wing-up: On the drive logs are allowed to fill in coves and backwaters so that the main body of logs may pass through easily.

LIST OF PEOPLE INTERVIEWED

The oral sources for this study are of three basic types.

Tape-recorded interviews with thirteen people; untaped sessions with individuals who have been of special assistance providing information, photographs, and other sources; individuals and organizations contacted by Thomas Armstrong in the course of this project. They provided names of people, places, and also photographs relating to lumbering on the Saco River. We thank them all for making this book possible.

I. Tape-recorded Interviews

For each person, the following information is given: name, address, date of birth, experience with J.G. Deering & Son on the Saco. Northeast Archives accession number: length of interview, pages of complete catalog, date(s) of interview(s). Dates of birth have sometimes been inferred from the interviews. The address given indicates where the informant lived while associated with Deering. All of us involved in this project have deep respect and admiration for their good long memories, knowledge of skills now disappearing, and good humor.

Bell, Leo, Fryeburg, ME, b. 1889. Timber Cruiser, scaler, river driver, woods boss 1916-1960. NA 1404: 2 hrs., 42 pp. catalog; September 15, 1980, November 17, 1980.

Bennett, Ralston, Lovell, ME, b. 1916. Trucker, woodsman 1940-1956. NA 1423: 30 min., 15 pp. catalog; November 17, 1980.

Blaney, Ken, Cornish, ME, b. 1910. Scaler 1934-1940's. NA 1403: 1 hr. 15 min., 34 pp. catalog; September 8, 1980.

Brown, Clarence, South Standish, ME, b. 1904. River driver, woodsman 1920's. NA 1406: 1 hr., 26 pp. catalog; September 16, 1980.

Burrill, Ed, Cornish, ME, b. 1901. River driver, woodsman 1929-1940's. NA 1405: 1 hr. 50 min., 44 pp. catalog; September 15, 1980, November 18, 1980.

Coker, Alan, Old Orchard Beach, ME, b. 1917 (?). Mill worker and manager at Diamond Match Company 1934-1979. NA 2195: 45 min., 8 pp. catalog; June 8, 1990.

Deering, Joseph G., Saco, ME, b. 1894. Manager and owner, J.G. Deering & Son 1920-1958. NA 1402: 2 hrs. 20 min., 55 pp. catalog; August 14, 1980, August 28, 1980.

Deschambeault, Phyllis, Saco, ME, b. 1910. Secretary for J.G. Deering & Son 1929-1958. NA 2194: 40 min., 5 pp. catalog; September 1, 1989.

Foran, Charles, Saco, ME, b. 1893. Retail sales, trucker, mill and yard foreman 1920-1950. NA 1407: 2 hrs., 83 pp. catalog; September 16, 1980, November 19, 1980.

Leonard, Chet, Fryeburg, ME, b. 1908. Deering portable mill foreman 1952-53. NA 1426: 40 min., 20 pp. catalog; November 18, 1980.

Littlefield, Robert, Lovell, ME, b. 1903. River driver, woodsman 1920-1930's. NA 1425: 1 hr., 27 pp. catalog; November 18, 1980.

Morin, Ralph, Saco, ME, b. 1918. Woodsman, J.G. Deering mill crew, retail salesman 1934-1981. NA 2010: 1 hr. 20 min., 43 pp. catalog; October 13, 1987.

Smith, Leroy, Lovell, ME, b. 1908. River driver, woodsman 1930's-1940's. NA 1424: 1 hr., 25 pp. catalog; November 18, 1980.

II. Individuals giving special assistance (untaped sessions)

These people provided special assistance in a variety of ways. They gave generously of their time and personal recollections, as well as photographs, artifacts, and written materials. They provided specific information on river landings, boom locations, and drive activities.

Mrs. Betty (Bell) Baker, Fryeburg
Mrs. Arlene Blaney, Cornish
Mrs. Arlene Chappell, Biddeford
Mr. & Mrs. Rudolph Danis, Saco
Mr. & Mrs. Robert Littlefield, Lovell
Mrs. Irene C. Maher, Biddeford
Mr. & Mrs. Carroll Perkins, Cornish
Mrs. Doris (Cunningham) Reed, Bass Harbor
Mr. & Mrs. Leroy Smith, Lovell
Rodney Warren, Steep Falls
Dyer Library and York Institute Museum, Saco (Emerson Baker, Director; Kerry O'Brien, Curator)
McArthur Library, Biddeford (Robert Filgate, Director; Jack Chisholm, Research)

III. Individuals and organizations contacted by Thomas M. Armstrong

The people listed below gave helpful suggestions and provided many leads, information and photos on lumbering in the Saco River Valley.

Philip G. Andrews, Fryeburg
Donald Anderson, Hollis
Mr. & Mrs. William Berry, Hollis
Mrs. Frances (Boulter) Hartford, West Buxton
Donald Boulter, Standish
Biddeford Historical Society, Biddeford
Donald Buzzell, Fryeburg
Francis G. Buzzell, Fryeburg
Mr. & Mrs. Kenneth Chaplin, Steep Falls
Mrs. John M. Chandler, Lovell
Mr. & Mrs. Wilfred H. Charron, Sebago Lake
Hubert Clemmons, Hiram
Mr. & Mrs. Edmund A. Choroszy, Saco
Raymond C. Cotton, Hiram
Mrs. Jean Cousins, Sebago Lake
Lee M. Cheney, Saco
Central Maine Power Co., Augusta
 (Judy L. Franke, Archivist)
Philip Dearborn, Limington
Reynald A. Eon, Biddeford
Roy P. Fairfield, Biddeford
Arnold G. Foster, Limington
Arthur L. Fournier, Saco
Harry Friedman, Old Orchard
Fryeburg Historical Society, Fryeburg
Fryeburg Fair Farm Museum, Fryeburg
Forest History Society, Durham, NC
John R. Fox, Lovell
Harold Gain, Fryeburg
Norman Gray, Fryeburg
Great Ossipee Museum, Hiram
Wilbur Hammond, Hiram
Thomas Hammond, Hiram
David Hastings, Sr., Fryeburg
Hugh Hastings, Fryeburg
Mrs. Virginia Henderson, Saco
Mrs. Catherine Dearborn Hilton, Limington
Mr. & Mrs. Freeman Howard, Hiram
Hollis-Buxton Historical Society
John Hubbard, E. Limington
Mrs. Gilberta (Usher) Ingalls, Alfred
Martin Jewett, West Buxton
Mr. & Mrs. Edward Jones, Fryeburg
Mr. & Mrs. Clayton King, Saco
Paul Lamontagne, Saco
Alexander C. Laroche, Biddeford
Robert Larry, Hollis Center
Barry Leighton, Hollis Center
Arthur Libby, South Standish
Limington Historical Society, Limington
Edward Lord, Limington
Steven E. Nichols, West Buxton
Mrs. Kenneth Noyes, Falmouth
Sumner Oleson, Westbrook
Francis M. O'Brien, Portland
Mrs. Anne (Whitehurst) Ordway, Freeport
Ancil Perkins, Standish
Mr. & Mrs. Richard Pinkos, Cornish
Mrs. Evelyn Pendexter, Steep Falls
Richard A. Pinkham, Gorham
Eugene R. Poston, South Portland
Paul Rivard, North Andover, Massachusetts
Winthrop Smith, Gorham
Sidney Stackpole, Saco
Mrs. Constance Sullivan, Saco
Robert C. Taylor, Auburn, ME
Harry L. Taylor, Steep Falls
Merton Towle, Hollis Center
Patricia (Foran) Tibbetts
Mr. & Mrs. David Tripp, West Buxton
Roland K. Usher, Hollis Center
Steve Usher, Gorham
Joseph M. Wagner, Alfred
Erwin Warren, Saco
Avon Wilcox, Alfred
Mrs. Albert E. Whitehurst, Saco
Mr. & Mrs. Welch, Brownfield
Clinton G. Woodsome, Waterboro
Robert A. Yarumian, West Buxton

BIBLIOGRAPHY

Anderson, Hayden L.V. *Canals and Inland Waterways of Maine*. Portland: Maine Historical Society, 1982.

Butler, Joyce. *Wildfire Loose*. Kennebunkport: Durell Publications, 1978.

Ballew, Stephen; Brooks, Joan; Brotz, Dona; and Ives, Edward D. *"Suthin'" An Oral History of Grover Morrison's Woods Operation at Little Musquash Lake, 1945–1947*. Orono, Maine: Northeast Folklore Society. *Northeast Folklore* XVIII: 1977.

Deering Papers. All of the correspondence, business records, and reports cited in this book are on file at the Dyer Library-York Institute in Saco, Maine. Copies are also on file at the Northeast Archives of Folklore and Oral History, Orono, Maine.

Dorson, Richard M., ed. *Handbook of American Folklore*. Bloomington: Indiana University Press, 1983.

Drake, Samuel, ed. *The Indian Wars of New England*. New York: Burt Franklin, 1971 (originally published in 1677).

Fairfield, Roy P. *Sand, Spindles, and Steeples: A History of Saco, Maine*. Portland: House of Falmouth, 1956.

Folsom, George. *History of Saco and Biddeford*. Saco: Alex C. Putnam, 1830.

Hilton, C. Max. *Rough Pulpwood Operating in Northwestern Maine 1935-1940*. University of Maine Studies, Second Series, No. 57. Orono: The University Press, 1942.

History of York County, Maine. Philadelphia: Everts and Peck, 1880.

Hubbard, William. "A Narrative of the Troubles with the Indians in New England from Piscataqua to Pemaquid." In *The Indian Wars of New England*, ed., Samuel Drake, pp. 105-110. New York: Burt Franklin, 1971.

Ives, Edward D. *Argyle Boom*. Orono, Maine: Northeast Folklore Society. *Northeast Folklore* XVII: 1976.

Ives, Edward D. *Joe Scott: The Woodsman Songmaker*. Urbana: University of Illinois Press, 1978.

Jewett, Martin H. and Hannaford, Olive W. *A History of Hollis, Maine 1660-1976*. Farmington: The Knowlton and McLeary Co., 1976.

Jones, Susan D. *Deering Family of Southern Maine*. Privately printed, 1979 (based on an earlier work by William E. Emery).

Libby, Charles T., ed. *Province and Court Records of Maine Vol. I*. Portland: Maine Historical Society, 1928.

Lindholt, Paul J., ed. *John Josselyn, Colonial Traveler: A Critical Edition of Two Voyages to New England*. Hanover: University Press of New England, 1988.

Maine Democrat. March 25, 1851.

Mitchell, Roger E. *I'm a Man That Works: The Biography of Don Mitchell of Merrill, Maine*. Orono, Maine: Northeast Folklore Society. *Northeast Folklore* XIX: 1978.

Mitchell, Roger E. "Occupational Folklore: The Outdoor Industries." In *Handbook of American Folklore*, ed. Richard M. Dorson, pp. 128-135. Bloomington: Indiana University Press, 1983.

Noyes, Sybil; Libby, Charles T. and Davis, Walter G., eds. *Genealogical Dictionary of Maine and New Hampshire 1928-1939*. Baltimore: Genealogical Publishing Co., 1979 (reprinted).

Pike, Robert E. *Tall Trees, Tough Men*. New York: W.W. Norton, 1967.

Rolde, Neil. *Sir William Pepperrell of Colonial New England*. Brunswick: Harpswell Press, 1982.

Sewall, James W. "Report on White Pine Tributary and Adjacent to the Saco River Valley in Maine and New Hampshire, 1943, for Mr. Joseph G. Deering." Located in Deering Papers, Dyer Library-York Institute, Saco, Maine.

Smith, David C. *A History of Lumbering in Maine 1861-1960*. University of Maine Studies No. 93. Orono: University of Maine Press, 1972.

Sullivan, James. *History of the District of Maine*. Boston: Thomas and Andrews, 1795.

Taylor, Robert C. *A History of Limington, Maine, 1668-1900*. Oxford: Oxford Hills Press, 1975.

Varney, George. *A Gazetteer of the State of Maine*. Boston: B.B. Russell, 1881.

Wells, Walter. *Water Power of Maine*. Augusta: State of Maine, 1867.

Wood, Richard G. *A History of Lumbering in Maine 1820-1861*. University of Maine Studies, Second Series, No. 33. Orono: The University Press, 1935. Reprinted 1972 (with Introduction by David C. Smith).

ADDENDUM

The following material has relevance to the sawmill diagram on page 57: The usual work force for the sawmill included one or two men working at the log slip on the river and fourteen men on the mill floor (triangles indicate the approximate location of each work station). In addition, four to six men worked at the green lumber platform or "brow," and there were one or two millwright/blacksmiths, one or two saw filers, and two to four teamsters or truckers.